Engineering Cyber-Physical Systems and Critical Infrastructures

Volume 5

Series Editor

Fatos Xhafa, Departament de Ciències de la Computació, Technical University of Catalonia, Barcelona, Spain

The aim of this book series is to present state of the art studies, research and best engineering practices, real-world applications and real-world case studies for the risks, security, and reliability of critical infrastructure systems and Cyber-Physical Systems. Volumes of this book series will cover modelling, analysis, frameworks, digital twin simulations of risks, failures and vulnerabilities of cyber critical infrastructures as well as will provide ICT approaches to ensure protection and avoid disruption of vital fields such as economy, utility supplies networks, telecommunications, transports, etc. in the everyday life of citizens. The intertwine of cyber and real nature of critical infrastructures will be analyzed and challenges of risks, security, and reliability of critical infrastructure systems will be revealed. Computational intelligence provided by sensing and processing through the whole spectrum of Cloud-to-thing continuum technologies will be the basis for real-time detection of risks, threats, anomalies, etc. in cyber critical infrastructures and will prompt for human and automated protection actions. Finally, studies and recommendations to policy makers, managers, local and governmental administrations and global international organizations will be sought.

Ryoichi Shinkuma · Fatos Xhafa · Takayuki Nishio
Editors

Advances in Engineering and Information Science Toward Smart City and Beyond

Springer

Editors
Ryoichi Shinkuma
Computer Science and Engineering
Shibaura Institute of Technology
Tokyo, Japan

Fatos Xhafa
Departament de Ciències de la Computació
Universitat Politècnica de Catalunya
Barcelona, Spain

Takayuki Nishio
School of Engineering
Tokyo Institute of Technology
Tokyo, Japan

ISSN 2731-5002 ISSN 2731-5010 (electronic)
Engineering Cyber-Physical Systems and Critical Infrastructures
ISBN 978-3-031-29300-9 ISBN 978-3-031-29301-6 (eBook)
https://doi.org/10.1007/978-3-031-29301-6

This Springer imprint is published by the registered company Springer Nature Switzerland AG
The registered company address is: Gewerbestrasse 11, 6330 Cham, Switzerland

Preface

Smart cities have emerged as highly complex technological endeavors that combine technology from many disciplines ranging from engineering to information science. This presents a multitude of challenges in various domains, including technical, economic, and social, across application subdomains of smart cities such as smart transportation, the Internet of Things, social welfare, tourism, and smart industry. The field covers a wide variety of topics as listed below (and not limited to):

Smart city pilot projects;
Prototyping for smart city applications and services;
Proof of concept of technologies enabling smart cities;
Computing/Networking/Sensing/Management frameworks and systems for smart cities;
Practical AI/Machine learning for smart cities;
Robust smart cities;
Security and Privacy for smart cities;
Autonomous mobility;
Robotics for smart cities;
Applications of unmanned aerial vehicles;
Blockchain technology for IoT and AI;
Green technology for sustainable cities;
Human-robot interaction/human in the loop;
System of systems/Meta system design for smart cities.

It becomes difficult or sometimes impossible to provide a solution for potential research issues contained in the above topics from only a traditional disciplinary approach; to tackle such research issues and to make the paradigm of smart cities a reality, interdisciplinary approaches are necessary. However, most of the studies published as academic works before are conceptual, theoretical, or simulation-based, and only a limited number have demonstrated feasibility through prototyping and proof of concept experiments.

The goal of this book is to present a bundle of works on the proof of concept with a prototype for smart cities and beyond from all the fields of engineering, such as architecture/civil engineering, materials, applied physics, applied chemistry, machinery, electrical/electronics, and information and all fields of information science such as Mathematical/Intelligent/Bio/Human/Social informatics. For experts in the fields of engineering and information science, the context of smart cities is a great opportunity to prototype their work in the real-life field, beyond the confines of laboratory experiments.

The editors received full-paper submissions only that describe previously unpublished, original, state-of-the-art research and that are not currently under review by a conference or journal. Accepted chapters were selected from the submissions through a peer-review process. As a result, this book volume brings readers 11 chapters with contributions to smart cities, which are arranged and summarized as follows.

The first chapter, "Critical Infrastructures Resilience in the Context of a Physical Protection System" by Rehak, presents the concept of assessing the resiliency of critical infrastructure elements by investigating its physical protection system. The essence of this concept is the definition of security factors of the physical protection system and the procedure for their use in assessing the critical infrastructures resilience against intentional physical threats. Using this tool, security managers of critical infrastructure entities can assess primary risks disrupting the critical infrastructure elements' physical protection system. Based on the information obtained, they can then identify the weak points of the protection system as well as the infrastructure itself and start the process of strengthening the critical infrastructure elements' resilience.

The second chapter, "Smart City Security Based on a Meta-Security Framework for Digital Twins" by Sugiyama, introduces a meta-security framework that spans cyberspace and physical space as a novel security measure for smart cities using digital twins. Security measures in the context of smart cities have been taken mainly in cyberspace; these measures include alerting users based on cyber threat information and observed events on the network. As the integration of cyberspace and physical space progresses toward beyond the 5th generation (B5G) era, attacks and their effects will expand not only to cyberspace but also to physical space, and it is expected that completely new security threats will emerge. The meta-security framework presented in this chapter is intended to offer comprehensive security solutions to oversee and harmonize multiple security mechanisms in both spaces.

The third chapter, "On-Premise Artificial Intelligence as a Service for Small and Medium Size Setups" by Fortuna, discusses AI as a Service (AIaaS) functionality and the corresponding technology stack. With new services introduced to smart cities and the increasing amount of data collected thereof, the in-time decision support will require the transition from developing a mostly static AIaaS approach. Various commercial solutions, typically deployed in a public cloud, already offer user-friendly and easy-to-use AIaaS, functionality-wise enabling the democratization of such ecosystems. However, open-source equivalent ecosystems supporting private deployment for better control over data, processes, and costs, where this is needed, are lagging behind, mostly because they require different sets of advanced skills to select and integrate the most suitable technologies for a given application area. This chapter analyzes possible realizations using open-source user-friendly technologies that are suitable for on-premise small and medium-sized setups allowing full control over the data and technology platform without any third-party dependence or vendor lock-in.

The fourth chapter, "Incentives in Surplus Food Distribution for Smart Cities and Beyond: An Activity Aware Solution" by Sanyal, discusses an incentive module that attracts volunteer participation and ensures their constant availability for the smooth movement of surplus food. Food surplus donation has emerged as a popular means to counter the food wastage issue that the world has been facing in recent years. The fact that this also addresses the food insecurity challenge is an added bonus. The matching of the surplus donors with the receivers needs the constant availability of willing volunteers to make this movement a success. While the existing incentive mechanism gets the attention of the volunteers, it fails to account for the degree of activity of the volunteers. This work modifies the incentive scheme to ensure that the activities of the volunteers are taken into consideration. The proposed mechanism manages to discourage inactivity with lower incentives and promote proactivity with higher incentives while disbursing the same total incentive amount taking all volunteers together.

The fifth chapter, "Evaluation of Smart Charging Integrated with Smart Energy Management and Advance Booking in an eMobility Urban Living Lab" by Jiang, discusses digital support for smart and green charging in an urban living lab setting, which has been proposed, demonstrated, and evaluated by the GreenCharge project. The proposed solutions are documented in a reference architecture meant to act as a blueprint both facilitating the extension and integration of the involved systems in the prototype implementation and supporting replication. However, successful uptake also depends on the implementation process, among others, the successful motivation and engagement of relevant stakeholders. Thus, the evaluation addressed in this chapter is based on the CIVITAS process evaluation methodology carried out within the context of design science and urban living labs. They share their experience and lessons learned from the design, implementation, and deployment of the proposed solutions in an urban housing cooperative in Norway, focusing on smart charging integrated with smart energy management and advance booking of charge points.

The sixth chapter, "Vehicle Allocation Algorithm Improving User Satisfaction in Ride-Sharing" by Wakazono, discusses ride-sharing services. To make ride-sharing services sustainable, maximization of user satisfaction is important. This chapter, therefore, proposes a vehicle allocation algorithm that can handle the individual user satisfaction preference. The key idea is to switch the cost function used in the optimization of vehicle allocation based on an individual user satisfaction preference. This work assumes that there are users whose satisfaction preference consists of quick arrival and economy. This work defines convenience, economy, and balance cost functions based on user satisfaction preferences of quick arrival and economy. Combining the cost functions with the existing successive best insertion (SBI) vehicle allocation algorithm, this work allocates vehicles to requests while maximizing user satisfaction.

The seventh chapter, "A City Airspace Testbed for Drone Networks in Future Smart Cities" by Yamazaki, proposes a city airspace testbed that can validate the feasibility of drone networks and services in real-world settings. To validate the feasibility of city services, service developers must implement the system in real-world settings such as urban areas. However, the experimenters cannot avoid various limitations and legal regulations. Their testbed has four key features: modularity, portability, safety, and agility for proof of concept in urban areas. The testbed comprises lightweight self-standing smart poles as the key modules that act as hovering drones rather than physical ones. This chapter demonstrates the three fundamental field experiment scenarios using the testbed.

The eighth chapter, "A Feasibility Study of Tethered Autonomous Moving Cells for Smart City" by Ota, discusses a tethered flying drone, which is becoming a promising option for next-generation mobile communications. According to the authors, tethered ground/floating nodes have not been employed for moving cells. This chapter proposes a concept of tethered autonomous moving cells for smart cities. The goal of the tethered moving cells is to provide stable and high data rate communication via optical fiber cables for unstable demands. The tethered cables also provide a long battery lifetime. This work also proposes the slackness control algorithm to optimally handle the optical cable. The feasibility of the proposed idea is demonstrated with the experimental results using a ground robot and a floating node.

The ninth chapter, "Relationship Among Different Types of Input and Model Accuracies in LSTM Driver Models" by Honda, aims to investigate the relationship between model inputs and model accuracy using a Long Short-Term Memory (LSTM) driver model in a car-following situation. The inputs to the model are the ego vehicle velocity, headway distance, relative velocity, log headway distance, Time Headway (THW), and Time to Collision (TTC). Six input patterns are set up to investigate the relationship between the inputs to the model and the model accuracies. In this work, the experiment is conducted using a driving simulator to simulate traffic congestion on a straight highway and measure the driver's driving data. The results show that THW is able to construct a more accurate driver model than TTC, and the model accuracy is improved when TTC and THW are used in combination as inputs: the headway distance for TTC and relative velocity for THW.

The tenth chapter, "Watch-from-Inside: 3D Sensor System Deployed Inside Building for Road Safety" by Ozawa, proposes a watch-from-inside system with 3D sensors deployed inside a building for road safety. When 3D data at an intersection is acquired by a Light Detection and Ranging (LIDAR) sensor through a window between the intersection and the sensor, the window causes deterioration of 3D data caused by attenuation, reflection, and refraction of the laser, which results in degradation of the performance of monitoring moving objects at the intersection. Therefore, we need to take such deterioration into account when we train machine learning models for object detection. The proposed system constructs a conversion model for data obtained without a window. By applying the model to data acquired without a window, it can reflect the deterioration of points caused by the window to the applied data.

The last chapter, "Federated Learning with Client Selection in Resource-Uncertain Wireless Networks: Simulation and Proof of Concept Experiments" by Yoshida, discusses Federated Learning (FL) in the smart city scenario. Data-holder devices called clients are heterogeneous, and their computation and communication resources available for FL are uncertain and fluctuating. This chapter presents a (MAB)-based client selection method for FL, which does not require prior knowledge of the computation and communication resources of clients. The MAB-based method achieves a better tradeoff between exploration and exploitation and reduces the time consumption for model updating. This chapter demonstrates that the proposed scheme requires less learning time than that required by the conventional method in a resource-fluctuating scenario. Experimental results using the off-the-shelf devices demonstrated the feasibility of FL and MAB-based client selection in a real wireless network environment.

As we conclude this preface, we would like to thank the authors for their contributions to the book. We appreciate the reviewers for their volunteer time and their constructive feedback to the authors to improve their chapter. The support from the Springer team is highly appreciated.

We hope readers will find the content of the book interesting and inspiring for their activity!

Tokyo, Japan — Ryoichi Shinkuma
Barcelona, Spain — Fatos Xhafa
Tokyo, Japan — Takayuki Nishio

Acknowledgments

Professor Ryoichi Shinkuma's work is supported in part by JSPS KAKENHI, Japan, under Grant no. 21H03427 and by the National Institute of Information and Communications Technology (NICT), Japan (No. 05201).

Professor Fatos Xhafa's work is supported by the research project of the *Ajuntament of Barcelona and the Fundació "laCaixa"* under the framework of the Barcelona Science Plan 2020–2023 (grant 21S09355-001).

Professor Takayuki Nishio's work is supported in part by JST, PRESTO Grant Number JPMJPR2035, Japan.

Contents

Contributors

Gregor Cerar Jozef Stefan Institute, Ljubljana, Slovenia

Carolina Fortuna Jozef Stefan Institute, Ljubljana, Slovenia

Svein Hallsteinsen SINTEF Digital, Trondheim, Norway

Ivo Haring Fraunhofer Ernst-Mach-Institut, EMI, Freiburg, Germany

Toshiya Hirose Department of Engineering Science and Mechanics, Shibaura Institute of Technology, Koto-ku, Tokyo, Japan

Kaisei Honda Human Machine System Laboratory, Shibaura Institute of Technology, Koto-ku, Tokyo, Japan

Martin Hromada Faculty of Applied Informatics, Tomas Bata University in Zin, Zlin, Czech Republic

Shigemi Ishida School of Systems Information Science, Future University Hakodate, Hokkaido, Japan

Shanshan Jiang SINTEF Digital, Trondheim, Norway

Panagiotis Kapsalis National Technical University of Athens, Attiki, Greece

Yasuaki Kobayashi KDDI Research, Saitama, Japan

Ayumu Kubota KDDI Research, Saitama, Japan

Vikash Kumar Singh School of Computer Science and Engineering, VIT-AP University, Amaravati, Andhra Pradesh, India

Karen Byskov Lindberg NTNU Electric Power Engineering, Trondheim, Norway; SINTEF Community, Oslo, Norway

Tomas Lovecek Faculty of Security Engineering, University of Zilina, Zilina, Slovak Republic

Sumiko Miyata Shibaura Institute of Technology, Tokyo, Japan

Takumi Miyoshi Shibaura Institute of Technology, Tokyo, Japan

Mihael Mohorčič Jozef Stefan Institute, Ljubljana, Slovenia

Masahiro Morikura Graduate School of Informatics, Kyoto University, Kyoto, Japan

Sajal Mukhopadhyay Department of Computer Science and Engineering, National Institute of Technology Durgapur, Durgapur, West Bengal, India

Din Mušić Jozef Stefan Institute, Ljubljana, Slovenia;
Jozef Stefan International Postgraduate School, Ljubljana, Slovenia

Masataka Nakahara KDDI Research, Saitama, Japan

Yu Nakayama Institute of Engineering, Tokyo University of Agriculture and Technology, Tokyo, Japan

Marit Natvig SINTEF Digital, Trondheim, Norway

Takayuki Nishio School of Engineering, Tokyo Institute of Technology, Tokyo, Japan

Shoko Oikawa Human Machine System Laboratory, Shibaura Institute of Technology, Koto-ku, Tokyo, Japan

Norihiro Okui KDDI Research, Saitama, Japan

Kaori Ota Department of Computer and Information Sciences, Tokyo University of Agriculture and Technology, Tokyo, Japan

Hajime Ozawa Shibaura Institute of Technology, Tokyo, Japan

David Rehak Faculty of Safety Engineering, VSB – Technical University of Ostrava, Ostrava, Czech Republic

Banhi Sanyal School of Computer Engineering, Kalinga Institute of Industrial Technology, Bhubaneswar, Odisha, India

Surja Sanyal Department of Computer Science and Engineering, Indian Institute of Technology Guwahati, Guwahati, Assam, India

Ryoichi Shinkuma Shibaura Institute of Technology, Tokyo, Japan;
Hyper Digital Twins Co., Ltd. (HDT), Tokyo, Japan

Yoh Shiraishi School of Systems Information Science, Future University Hakodate, Hokkaido, Japan

Keizo Sugiyama KDDI Research, Saitama, Japan

Yuta Wakazono Graduate School of Systems Information Science, Future University Hakodate, Hokkaido, Japan

Neil Walker International Association of CIP Professionals, London, UK

Hirofumi Watanabe SoftBank Corporation, Tokyo Portcity Takeshiba Office Tower 27F, Tokyo, Japan

Fatos Xhafa Department of Computer Science, Universitat Politècnica de Catalunya, Barcelona, Spain

Koji Yamamoto Graduate School of Informatics, Kyoto University, Kyoto, Japan

Taku Yamazaki Shibaura Institute of Technology, Tokyo, Japan

Naoya Yoshida Graduate School of Informatics, Kyoto University, Kyoto, Japan

Andrej Čampa Comsensus, Ljubljana, Slovenia

Critical Infrastructures Resilience in the Context of a Physical Protection System

David Rehak, Tomas Lovecek, Martin Hromada, Neil Walker, and Ivo Haring

Abstract Building the Smart City concept requires a high level of technical support, especially in the area of energy and ICT infrastructure. For this reason, these infrastructures must show a high level of resilience against the effects of current security threats. Currently, the most significant threats also include intentional threats of a physical nature, e.g. terrorist attacks. A very effective way to protect critical infrastructures against these threats is the implementation of a physical protection system. This system enables the integration of technical, organizational and regime (policy) measures preventing the effect of intentional physical threats. Based on these facts, the chapter presents the concept of assessing the critical infrastructures elements resilience by investigating its physical protection system. The essence of this concept is the definition of security factors of the physical protection system and the procedure for their use in assessing the critical infrastructures elements resilience against intentional physical threats. By using this tool, security managers of critical infrastructure entities can carry out an assessment of primary risks disrupting the critical infrastructure elements' physical protection system. Based on the information obtained, they can then identify the weak points of the protection system as

D. Rehak (✉)
Faculty of Safety Engineering, VSB – Technical University of Ostrava, Ostrava, Czech Republic
e-mail: david.rehak@vsb.cz

T. Lovecek
Faculty of Security Engineering, University of Zilina, Zilina, Slovak Republic
e-mail: tomas.lovecek@uniza.sk

M. Hromada
Faculty of Applied Informatics, Tomas Bata University in Zin, Zlin, Czech Republic
e-mail: hromada@utb.cz

N. Walker
International Association of CIP Professionals, London, UK
e-mail: neilw@torchmarketing.co.uk

I. Haring
Fraunhofer Ernst-Mach-Institut, EMI, Freiburg, Germany
e-mail: haering@emi.fraunhofer.de

R. Shinkuma et al. (eds.), *Advances in Engineering and Information Science Toward Smart City and Beyond*, Engineering Cyber-Physical Systems and Critical Infrastructures 5, https://doi.org/10.1007/978-3-031-29301-6_1

well as the infrastructure itself and start the process of strengthening the critical infrastructure elements resilience.

Keywords Critical infrastructures · Resilience · Physical protection system · Smart cities · Energy · ICT

List of Acronyms

ASD	Adversary Sequence Diagram.
ASOR	Assessing and Strengthening Organisational Resilience.
ASSESS	Analytical System and Software for Evaluating Safeguards and Security.
CDP	Critical Detection Point.
CIERA	Critical Infrastructure Elements Resilience Assessment.
CISI	Critical Infrastructure Spatial Index.
EACS	Electronic Access Control System.
EAS	Emergency Alarm System.
EASI	Estimate of Adversary Sequence Interruption.
ESS	Electronic Security System.
EU	European Union.
FTA	Fault Tree Analysis.
ICT	Information and Communication Technology.
PESTLE	Political, Economic, Social, Technological, Legal, Environmental.
PPS	Physical Protection System.
RMI	Resilience Measurement Index.
SAPE	Systematic Analysis of Physical Protection Effectiveness.
SATANO	Security Assessment of Terrorist Attack in a Network of Objects.
SAVI	Systematic Analysis of Vulnerability to Intrusion.
VSS	Video Surveillance System.

Glossary of Most Important Terms

Critical infrastructure	An asset or system which is essential for the maintenance of vital societal functions, health, safety, security, economic or social well-being of people.

Resilience	The ability to reduce the magnitude and/or duration of disruptive events. The effectiveness of a resilient infrastructure or enterprise depends upon its ability to anticipate, absorb, adapt to, and/or rapidly recover from a potentially disruptive event.
Resistance	The ability of a critical infrastructure element to prevent the occurrence of a disruptive event.
Robustness	The ability of a critical infrastructure element to absorb the impact of a disruptive event that has already occurred.
Physical protection system	Integrates people, procedures, and equipment for the protection of assets or facilities against theft, sabotage, or other malevolent human attacks.
Physical resistance	The ability to withstand the negative effects of naturogenic, anthropogenic and technogenic threats.
Security measures	A set of technical and organizational measures for both monitoring and physical protection of elements.
Detection capability	The probability and/or time of detection of a disruptive event.
Responsiveness	The probability and/or time of intervention leading to the elimination of the causes of a disruptive event or the minimisation of its consequences.

1 Introduction

Modern cities are currently completely dependent on the services provided by physical and socio- economic infrastructures [1]. There is considerable interdependence between the two infrastructure groups [2]. This can also be formulated in terms of infrastructure functions and services provided, independent of the technologies used [3]. On the one hand, almost all socio-economic infrastructures require unlimited availability of basic physical infrastructure services, e.g. energy or information and communication technology (ICT). And on the other hand, in the event of a crisis, basic physical infrastructures are dependent on the services of socio-economic infrastructures, e.g. the functioning of emergency services. Moreover, some of these infrastructures are so important that their disruption or failure would have a major impact on the interests protected by the state, i.e. the security of the state, the economy and the basic life needs of the population [4]. These infrastructures are therefore referred to as critical and together form a critical infrastructure system.

The Smart City concept currently enables the interconnection of various interdependent urban infrastructures [5]. However, even this concept itself is primarily dependent on the ICT infrastructure [6], which is further dependent on the supply of electricity. For this reason, it can be stated that energy and ICT infrastructures are unequivocally key in the Smart City concept, and therefore it is necessary to ensure their resilience against the occurrence and impact of disruptive events in the long term. Note that the smart city concept is somewhat focussing on local areas and related supply grids, e.g. community electricity distribution grids [7], as opposed to gas transmission grids on European scale [8].

The basic building block of energy and ICT infrastructures are buildings, devices or systems that are collectively referred to as critical infrastructure elements [4]. Specifically, the elements of these two infrastructures are permanently exposed to various security threats with varying degrees of intensity. However, in the context of the current security environment, meteorological phenomena [9] and terrorism [10] can be considered the most serious threats. Despite the fact that meteorological threats in some cases reach a very high level of intensity, it is currently possible to predict them in time, which enables the adoption of adequate security measures. However, in contrast, terrorist attacks can only be predicted to a very limited extent [11–13].

This fact is evidenced by a U.S. study. National Academy of Sciences [10], in which three key areas of the electricity supply system that show the greatest vulnerability to terrorist attacks are presented. These areas are physical vulnerability, cyber vulnerability and personnel vulnerability. For this reason, it is necessary to ensure a sufficient resilience level against intentional threats of a physical nature for these three areas. A suitable solution in this case could be the implementation of a physical protection system [14], which could be an important factor in strengthening the critical infrastructure elements resilience. As there is currently no approach linking these two research areas, the aim of this chapter is to try to integrate them and expand the foundations for further follow-up research.

2 Perceiving the Resilience in a Critical Infrastructure System

The term resilience was first defined by Holling [15] in connection with the resistance and stabilization of ecological systems (later also socio-ecological systems). Over time, the concept of resilience began to be reflected in other scientific fields such as sociology, psychology or economics. The relatively youngest field, from the point of view of investigating the resilience of systems, is the engineering domain [16]. In the context of critical infrastructure, resilience represents the internal preparedness of subsystems for disruptive events. It is thus the ability of these subsystems to ensure and maintain their functions under the negative influence of internal and/or external factors. Resilience can therefore be understood as the opposite of vulnerability, or resilience and vulnerability have an inverse character to each other. Vulnerable subsystems lack resilience and conversely resilient subsystems are not very vulnerable.

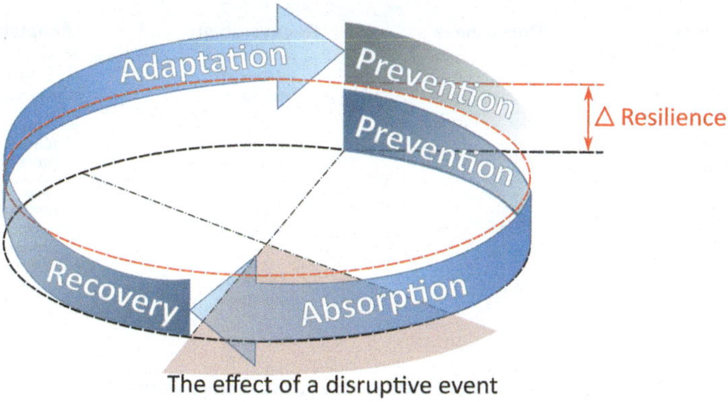

Fig. 1 Resilience cycle in a critical infrastructure system [18]

Resilience in the context of critical infrastructure was first described in the document Critical Infrastructure Resilience Final Report and Recommendations [17], where it is defined as "the ability to reduce the magnitude and/or duration of disruptive events. The effectiveness of a resilient infrastructure or enterprise depends upon its ability to anticipate, absorb, adapt to, and/or rapidly recover from a potentially disruptive event. "In connection with this definition, it is necessary to perceive resilience in a critical infrastructure system as a cyclical process of continuous improvement of prevention, absorption, restoration and adaptation of individual elements of critical infrastructure. One cycle is presented in Fig. 1 within which resilience is strengthened from the original level (i.e. black dashed line) to a new level (i.e. red dashed line). The difference between these Δ levels is understood as a measure of resilience enhancement.

The resilience cycle presented in Fig. 1 is divided into four phases. Theses phases can be understood as being ordered in time, namely ordered before, during and after managing disruption events, besides their interpretation as related resilience assessment aspects to be considered, e.g. [19].

The first phase of the cycle is prevention, which is determined by the resistance of critical infrastructure elements. By implementing preventive activities, the owner/operator prepares elements for future disruptive events; preparedness is therefore the resulting state of prevention. At the time of the action of such events, resilience passes from the phase of prevention to the phase of absorption. The second phase of the cycle is absorption, which is initiated by the influence of a disruptive event and is determined by the robustness of critical infrastructure elements. The essence of robustness is the ability of critical infrastructure elements to absorb the impact of a disruptive event with as less as possible fluctuations in the services provided by them.

After the disruptive event has ended, the third phase of the cycle occurs, which is recovery. It is characterized by recoverability, i.e. the ability of the elements to restore their activity to the original, or required performance levels. The length of

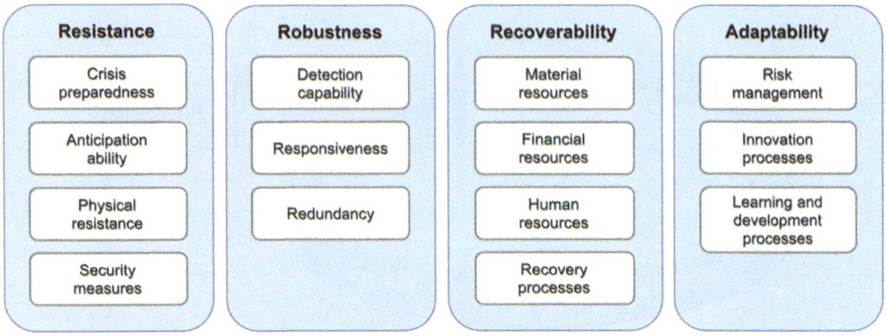

Fig. 2 Variables determining the resilience of critical infrastructure elements (modified from [22])

the recovery phase is determined by the available resources and the time required to implement individual recovery processes. The final phase of the critical infrastructure resilience cycle is adaptation. This is the ability of the subject of the critical infrastructure to adapt the operating element to the possible repetition of a disruptive event that has already taken place—that is, to learn from disruptive events solved in the past. Adaptation thus represents a dynamic, long-term, ability of the subject to adapt its elements to a changed situation.

Based on the above, the resilience of critical infrastructure elements can be understood as a state that is formed by four characteristic components, namely resistance, robustness, recoverability and adaptability. These four components or criteria of resilience are further determined by individual variables, which are presented in Fig. 2. Note, that we do not distinguish between these capabilities or phases and their analytical assessment, for instance as conducted in an explicit panarchy (infinity sign like) process that combines the resilience cycle and related capabilities with an analytical assessment process similar to ISO 31000 [20] using tabular or extended analytical system analysis methods [21].

Resistance is the ability of a critical infrastructure element to prevent the occurrence of a disruptive event. Its abilities as refined in Fig. 2 are understood as measures of a preventive nature that determine structural resistance and security resistance. They can be of physical, technical, cyber, organisational and to a lesser degree of social, psychological, regulative, legal, policy or political nature. These measures (variables) are:

- crisis preparedness (a set of analytical and planning documents to increase the elements' readiness for disruptive events and the implementation of subsequent security measures);
- anticipation ability (the ability to predict the possible occurrence of disruptive events);
- physical resistance (ability to resist the negative effects of natural, anthropogenic and technogenic threats);

- security measures (set of technical and organizational measures for monitoring and physical protection of elements).

Robustness is the ability of a critical infrastructure element to absorb the impact of a disruptive event that has already occurred. These impacts can be absorbed through early recognition and management of a disruptive event. Robustness is determined by the following variables (see Fig. 2):

- detection capability (probability and/or time of recognition of a disruptive event);
- responsiveness (probability and/or time of intervention leading to the elimination of the causes of a disruptive event or the minimization of its consequences);
- redundancy (the ability to immediately substitute the performance of a disturbed part of the element or strengthen its capacities).

Recoverability is the ability of a critical infrastructure element to restore its activity to the original (required) level of provided services after the effects of a disruptive event have ended. In the field of critical infrastructure, recoverability is understood as repairability; therefore, only the repair or replacement of damaged or destroyed components of the element is considered. Recoverability is determined by the following variables (see Fig. 2):

- material resources (availability of necessary components to carry out repair or replacement of damaged or destroyed parts of the element);
- financial resources (availability of financial resources, or reserves, enabling the financing of rapid restoration of the element);
- human resources (availability of human resources with the necessary qualifications);
- recovery processes (processes supporting rapid recovery of the element's required performance).

Adaptability is the ability of the critical infrastructure subject of the (i.e. the organization) to prepare the element for the repeated action of a disruptive event that has already taken place. It represents the dynamic (long-term) ability of the organization to adapt to a changed situation. Adaptability is determined by the internal processes of the organization, the essence of which is to create the best possible conditions for strengthening resistance, robustness and adaptability. The basic processes strengthening the adaptability of critical infrastructure elements to disruptive events are the following variables (see Fig. 2):

- risk management (processes supporting early risk assessment and management, including specification of scenarios of disruptive events);
- innovation processes (processes supporting invention, science and research and implementation of security measures);
- learning and development processes (processes supporting the knowledge, skills and attitudes of employees of the critical infrastructure entity).

The variables presented above, which determine the critical infrastructure elements resilience, are shaped by individual security factors. These factors come from a variety of areas, which can be classified as external and internal. The external areas of safety factors are divided into six areas corresponding to the PESTLE method [23], the essence of which is the analysis of external factors of the organization. These areas are political, economic, social, technological, legislative, and ecological. In contrast, the internal areas of security factors include four areas covering the main functional areas and management processes of the organization [24]. These areas are personnel, substantive, procedural and financial.

In the context of the above-defined vulnerability of critical infrastructure elements to intentional physical threats (e.g. terrorist attacks), the physical protection appears to be the most important area for the formation of security factors. Key security factors in this area can be identified primarily in the physical protection system [14], as its measures have the potential to shape several of the resilience variables presented above. These are mainly variables determining components of a preventive and absorptive nature, i.e. resistance and robustness. In the case of resistance, these are the variables "physical resistance" and "security measures". In the case of robustness, these are the variables "detection capability" and "responsiveness". The identification and more detailed description of the specific security factors of these variables are presented in Sect. 4 of this chapter.

3 Physical Protection of Critical Infrastructures Elements

In accordance with the Directive of the Council of the European Union [4], the critical infrastructures protection means all activities aimed at ensuring the functionality, continuity and integrity of critical infrastructures with the aim of preventing, mitigating and neutralizing threats, risks or vulnerabilities. In the context of physical protection, a system is termed effective that meets the basic condition that the time span between the point in time of an undesirable event detection (e.g. of a security event/incident) till the point in time where the damage effects of the intrusion damage are irreversible is longer than the reaction time of the intervention (reaction time shorter than intrusion complete time). This means that the system is functional when the ratio of these times in the given order is greater than one [25].

Credible proof of the fulfilment of this basic and seemingly elementary condition for the effectiveness of the protection system is, however, often difficult to achieve in practice. Existing procedures aimed at protecting critical infrastructure use two basic approaches, namely qualitative and quantitative. Procedures using a qualitative approach are based on expert assessors' estimation, where it is not possible to exactly verify the sufficiency of the protection proposed level and one must rely on the professional competence of these procedures' creators. In this case, it is not possible to verify whether the protection system from the proposed protective measures point of view is not under-dimensioned or, on the contrary, over-dimensioned. In contrast, procedures based on a quantitative approach enable the use of measurable inputs[1] and output quantities[2] accurately demonstrating the justification of the proposed protective measures. In this case, it is already possible to verify whether the protection system, from the proposed protective measures point of view, is not undersized or oversized.

Damage or destruction of critical infrastructures can be caused by intentional acts of terrorism, natural disasters, negligence, accidents, computer piracy or criminal activity and acts of bad faith. To protect the lives and property of the EU population against the risk of terrorism, natural disasters and accidents, any breach or manipulation of the most important infrastructures should be as short as possible, rare, manageable, geographically isolated and posing the least possible threat to the well-being of the Member States, their citizens and the European Union [26].

In 2005, the European Commission approved the "Green Paper on the Program for the Protection of Critical Infrastructure" [26], according to which these infrastructures include physical and cyber systems to ensure activities necessary for the functioning of the economy and state administration. Although this document lists possible methods/tools for improving prevention, protection, preparedness and response within the framework of the critical infrastructure protection in EU conditions, it does not further specify them. This is left to the implementation of individual member states within their legislative binding regulations. For example, according to the document Concept of critical infrastructure in the Slovak Republic and the method of its protection and defence [27], technical means for deterrence, detection, verification, signalling and the intruder elimination (mechanical and electronic) as well as the activity of security services, e.g. the intervention of security forces and armed forces.

In 2008, the European Council directive on the identification and designation of European critical infrastructures and the assessment of the need to improve their protection was adopted [4], which requires critical infrastructure elements operators to develop security plans and at the same time specifies the procedures for their development. Such a security plan identifies critical infrastructure facilities and security solutions that exist or are being implemented to protect them. At a minimum, the security plan should include:

---

[1] Examples for quantification of protection system properties: Breakthrough resistance times of mechanical barriers, total reaction time of the intervention unit, times of intruder movement, movement times of intervention units, intruder attack time, intervention time, alarm announcement time, attack verification time, probability of correct detection by the alarm system in the ith detection zone during the trip intruders, the probability of a fault-free state of the alarm system, the probability of a fault-free transmission of the alarm signal via the alarm transmission path to a remote alarm receiving centre, the probability of a timely and correct evaluation of the alarm state, etc.

[2] Examples for quantification of effectiveness of protection system: The coefficient of effectiveness of protective measures, the probability of eliminating the intruder, the cumulative probability of (correct) detection of the intruder, the probability of a successful intervention of the intervention unit, the critical value of the maximum interval between two inspections of the guard service, etc.

- identification of important devices,
- risk assessment based on the main threat scenarios, vulnerable points of individual devices and possible consequences,
- identification, selection, and prioritization of suitable protection measures.

For this purpose, it is possible to use, e.g., geospatial data which are fundamental for in-depth risk analyses of critical infrastructure elements. One such tool is the Critical Infrastructure Spatial Index (CISI) whose dataset can be applied to explore the current landscape of critical infrastructure, identify critical infrastructure hotspots, and as exposure input for large-scale risk assessments [28]. Another suitable tool is the dataset 'HARmonized grids of Critical Infrastructures in EUrope' [29]. HARCI-EU provides a consistent mapping of critical infrastructures in key sectors that can serve as exposure information for large-scale risk assessments in Europe.

When choosing protective measures, a distinction is made between permanent and graduated protective measures [4]. Permanent security measures specify the investments and resources necessary in the area of security that could be used at any time. This includes information on general measures such as technical measures (including the implementation of means of detection, access control, protection and notification), organizational measures (including procedures for warning and crisis management), control and verification measures, communication, awareness-raising and training, and security of information systems. Gradual protection measures can be activated depending on different levels of risk and threats.

Research in the field of a quantitative approach to the evaluation of the effectiveness of physical protection systems (PPS) dates back to the 1970s, when the first mathematical model for the Estimate of Adversary Sequence Interruption–EASI [30]. Ten years later, the method and the first software tool SAVI (Systematic Analysis of Vulnerability to Intrusion) was created, which applied the ASD (Adversary Sequence Diagram) method, enabling a graphical representation of possible attack paths in PPS [31, 32]. SAVI combines the EASI and ASD methods and evaluates all possible routes to the central zone in terms of probability of interruption and creates a list of the ten most vulnerable routes in terms of probability of interruption. The SAVI program is supplemented by an extensive database of delay and detection parameters of the most frequently used protective elements.

In 1989, the methodology and software tool ASSESS (Analytical System and Software for Evaluating Safeguards and Security) was created, which was an extension of the SAVI methodology, and which contains additional modules for analysing the destruction of the intruder and analysing the internal intruder and the cooperation of the internal and external intruder [33, 34].

The theory of a quantitative approach to the object protection systems assessment, which until then was mainly used in the USA to protect strategic objects such as nuclear and military facilities, was comprehensively described and at the same time promoted among the wider security community in the world by Garcia in his publication Design and Evaluation of Physical Protection Systems [25]. The publication of this publication can be considered an important milestone in the development of further research in the field of comprehensive assessment of PPS in the 21st century.

In 2008, the software tools SPRUT and Vega-2 were created to enable quantitative assessment of PPS. SPRUT is a software tool designed to assess the of nuclear facilities physical protection effectiveness developed by the Russian company ISTA. The software is used to model the combat match of intruders with physical protection. The newer version of SPRUT-IM uses simulation modelling of the penetration of intruders into the object with the aim of calculating the most effective scheme to thwart the attack [35]. Vega-2 is also a Russian software tool designed to determine the nuclear facilities physical protection systems effectiveness given the structure of the protection system and various intruder models (both internal and external). The Vega-2 software is used to create proposals and select the optimal option for improving the objects physical protection. The guarded area is modelled using zones, sections and transitions. The software includes a database of physical barriers and detection elements with the possibility of defining your own elements. The analysis also considers the activities of police officers and members of the intervention unit, and one of the outputs is a list of optimal routes during intervention [35].

In 2009, an article presenting a new approach and software tool, SAPE (Systematic Analysis of Physical Protection Effectiveness), was published, which builds on the SAVI and ASSESS methods, but significantly improves them. The method uses a new 2D guarded space model instead of the simple ASD method, and also uses a heuristic algorithm that substantially extends the sensitivity analysis, which analyses all protection measures on the most vulnerable path [36]. The resulting values then represent the relative effectiveness of updating individual protective measures. Less resolving is in comparison fault tree analysis (FTA) based physical intrusion assessments called attack trees, see [37].

In his work, Lovecek et al. were the first to define four basic theoretical models considering the decision-making of the intruder for certainty, or uncertainties, as well as the stochasticity of the input variables [38]. These models were subsequently used within the software tool SATANO (Security Assessment of Terrorist Attack in a Network of Objects), which enables quantitative assessment of PPS on 2D map data. The main innovative contribution of this software tool is the possibility of creating more complex attack scenarios in which the attack method/vector changes, such as an attack by a natural person using different types of tools [39]. Evaluation of PPS effectiveness of design and mitigation options using quantitative assessment is given in [40].

Currently, the objects physical protection systems assessment is understood in the broader context of the resilience assessment, which consists of four phases, i.e. prevention, absorption, restoration, adaptation [18]. From the point of view of the physical protection system, the first two phases are crucial in the resilience assessment, which are focused on measures of a preventive and absorptive nature at the time of the threat.

Prevention, as the first phase of the cycle of resilience, is determined by the resistance of critical infrastructure elements. The essence of resistance is the ability to prevent the occurrence of a disruptive event. The key variables of resistance are physical resistance in relation to security measures. Physical resistance is a set of mechanical barriers designed to increase the physical resistance of a critical infrastructure element to disruptive events. Security measures are understood as a set of technical means, physical protection elements and organizational measures.

Absorption, as the second phase of the resilience cycle, is initiated by the impact of a disruptive event and is determined by the robustness of the critical infrastructure element. The essence of robustness is the ability of a critical infrastructure element to absorb the impact of a disruptive event with as less as possible fluctuations in the main services provided by it. The key robustness variables are the element's detectability and responsiveness. Detection is understood as the probability and/or time of recognizing a disruptive event, while responsiveness is understood as the probability and/or time of intervention leading to the elimination of the causes of a disruptive event or the minimization of its consequences.

The common requirement of the components of resistance and robustness is to create a physical protection system that, as an efficient way of organizing measures, will prevent the occurrence of a disruptive event (i.e. repelling an intruder), or prevents the infringer from achieving his goal, which can be, for example, theft, damage or destruction of the protected interest (i.e. eliminating the infringing attack system or imprisoning an intruder). Such a system of physical protection can be understood as a system implemented by measures that can be divided into [25]:

- mechanical barriers,
- alarm systems,
- security forces,
- regime measures.

Mechanical barriers serve to deter, slow down, or stopping the intruder, while the alarm systems are used for its subsequent detection and triggering of the alarm state. An integral part of the protection system are the security forces, which ensure timely intervention and detention of the intruder. Regime protection ensures the proper functioning of the mentioned protection measures [41]. These protective measures must be arranged in such a way as to meet the basic requirement for the functionality of the system, namely that from the initial point of detection, the attack time is greater (including the total time of breaking the mechanical barriers and the time of the intruder's movements) than the response time of the intervention unit.

Mechanical barriers are formed by a system of natural and artificial physical obstacles (e.g. fences, building structures, opening fillings, safes, barriers, etc.), the primary purpose of which is to act on the intruder so that he abandons his intention. If they do not abandon their intention, then their task is to create a sufficient time delay between the time of attacking the protected area and the achievement of its goal, i.e. to make it difficult or practically impossible for the intruder to penetrate to the protected interest (e.g. key technologies, procedural regime at workplace, dispatch (separation, isolation) of a critical infrastructure element).

Each barrier is surmountable, depending on the amount of energy required, time, physical propositions/skills of the intruder and the type of tools used. Barriers have their basic physical, mechanical and insulating properties. Due to passive resistance, mechanical properties are key. In the case of an attempt to overcome, each of these means is stressed, for example in tension, pressure or shear. In real conditions, it is a combined stress, on the basis of which an accompanying phenomenon arises in the form of vibrations and acoustic perception. In addition to strength, barriers have another important property. This is a sound with different acoustic intensity, which is created by the vibration of the material, as a reaction to the resistance imposed by the barrier during violent action on its integrity using a certain tool. The resulting soundscape then becomes a very unpleasant accompanying phenomenon of an intrusion activity and can serve as an impulse to detect violations of the protected space, e.g. drilling, prying, cutting, demolition [42].

An alarm system is defined as an electrical installation that responds to manual or automatic detection of the presence of danger [43]. A similar definition of an alarm system can be found in the technical standard [44], where an alarm system is an electrical device that responds to a manual stimulus or to automatic detection of the presence of danger. However, this standard introduces the concept of an alarm application, which means an application designed to protect life, property or the environment. Such an application can be, for example:

- alarm security and cherry system,
- the call for help system,
- elevator alarm system,
- environmental impact alarm system,
- closed circuit television,
- input control system,
- electronic fire alarm.

In the case of the definition of an alarm application, its purpose is understood in a broader context than simply detecting the presence of danger in the form of an intruder. An example can be a closed television circuit, the purpose of which, in addition to detection, can also be monitoring, observation, recognition, identification, or investigation [45]. Non-alarm systems are systems intended only for control (e.g. heating, air conditioning, lighting, energy system control, building management) and their primary function is not the protection of life, property or the environment [44]. An integrated alarm system integrates multiple systems/applications, at least one of which is an alarm system [44].

The technical standard [46] distinguishes three types of alarm systems:

- alarm security system, an alarm system designed to detect and signal the presence, intrusion or attempted intrusion of an intruder into guarded premises,
- emergency alarm system, a system providing the user with the possibility of deliberately triggering an alarm condition,
- the call for help alarm system is a system providing the means to call for help and is intended for persons who may be considered as persons living in danger.

A whole set of technical standards is devoted to alarm systems, which generally divide alarm systems into:

- electronic security system / emergency alarm system (ESS/EAS) [47],
- video surveillance system (VSS) [45],
- electronic access control system (EACS) [48],
- emergency call system [49],
- alarm transmission system and equipment [43].

An integral part of any physical protection system are security forces that ensure timely intervention and, if necessary, restriction of personal freedom of the intruder. In professional literature and in legal regulations, the term physical protection means the performance of protection by the physical presence of persons in the protected area, or in its vicinity [41]. The most common persons who perform physical protection are members of security services or units, armed forces and members of the guards. In general, we can talk about members of the security services.

Security services can be divided according to several criteria [50]:

- relation to the state,
- the founder, or operator,
- characteristics of legal regulation,
- direction and type of activity of security services,
- territorial and substantive powers,
- levels of management and security forces.

The last subgroup of the physical protection system is operational and strategical regime measures. These are focused on the preparation and creation of procedures and processes that will ensure a functional framework for the effective and efficient application of the object's physical protection system. This also includes the activities of responsible workers in the sense of:

- assessing risk management,
- processing security documentation (e.g. strategies, policies, directives, projects, etc.),
- determining the necessary regime and organizational measures to ensure access to objects,
- determination of the working regime with classified facts, personal data and business, banking, or leaf secret (company confidential artefacts),
- audit and control activities.

The critical infrastructure protection is currently regulated by European and national generally binding legal regulations, which result in minimum requirements for protective measures. Procedures using a qualitative approach to the critical infrastructures protection are based on expert evaluators estimation, where it is not possible to exactly verify the sufficiency of the protection proposed level and one must rely on the professional competence of the creators of these procedures.

In contrast, procedures based on a quantitative approach to the critical infrastructures protection make it possible to accurately prove the validity of the proposed protection measures using measurable input and output variables. As already mentioned, the approaches of protection systems' quantitative assessments date back to the 1980s and were focused on assessments of military and energy facilities protection. Currently, the assessment of object protection systems is understood in the broader context of the assessment of the overall critical infrastructure system resilience. On the basis of the above, the following part of the chapter identifies physical protection factors that can be used to assess the critical infrastructure elements resilience, especially their resistance and robustness.

4 Security Factors of the Physical Protection System

On the basis of the detailed description of the critical infrastructures elements physical protection system, presented in the previous part of this chapter, it is possible to proceed to identify specific security factors of the physical protection system that shape individual variables of critical infrastructures elements resilience (see part 2 of this chapter). An overview of the identified security factors of the physical protection system is presented in Table 1.

The security factor of physical resistance is the breakthrough resistance of mechanical barriers (T_{MB}) expressed in units of time. This time indicates the ability of the mechanical barrier to resist an intruder's attack for a certain period of time and thereby to delay or to stop its progress to the protected interest. The physical resistance of a critical infrastructure element can then be understood as its overall property of resisting the efforts of an intruder. It is the sum of the breakthrough resistances of individual mechanical barriers on the path of the intruder to the protected interest. Breakthrough resistance is mainly influenced by the mechanical properties of the barrier materials used in relation to the type of tools used by the intruder. For evaluating the breakthrough resistance of Mechanical Barriers T_{MB} it applies that the

Table 1 Security factors of the physical protection system shaping the resilience variables of critical infrastructure elements

Resilience components	Resilience variables	Security factors of the physical protection system
Resistance (C_{RES})	Physical resistance (V_{PR})	Breakthrough resistance time of mechanical barriers (T_{MB})
	Security measures (V_{SM})	Security level of the alarm system (L_{AS})
		Preparedness and equipment level of the security forces (L_{SF})
		Level of regime measures standardization (L_{RM})
Robustness (C_{ROB})	Detection capability (V_{DC})	Cumulative probability of detection (P_{KDET})
		Critical detection point (T_{CDP})
	Responsiveness (V_{RN})	Probability of intruder elimination ($P_{R\|A}$)

Table 2 Security levels of the electronic security system (EES) and the emergency alarm system (EAS) [47]

Security level	Risk level	Intruder abilities
Level 1	Low risk	The intruder has little knowledge of ESS/EAS and has a limited range of available tools
Level 2	Low to medium risk	The intruder has some knowledge of ESS/EAS and has a basic assortment of tools and portable devices available
Level 3	Medium to high risk	The intruder is familiar with ESS/EAS and has a full range of tools and electronic equipment
Level 4	High risk	The intruder has the opportunity to prepare a detailed intrusion plan, and has a complete range of equipment, including the means to replace the ESS/EAS components. It is used when security takes priority over other considerations

total time of DELay of the intruder on the way to the protected interest T_{DEL} must be greater than the REAction time of the intervention unit T_{REA} [41].

In the field of security measures, three security factors have been identified regarding alarm systems, security forces and regime measures. The security factor of Alarm Systems is their level of security (L_{AS}), which expresses the degree of significance of the protected area in the context of risks, or skills and knowledge of the potential intruder. Depending on the classification of the alarm system in one of the listed levels, certain functional requirements are imposed on it in terms of reliability, resistance, access, probability of detection or special functions, e.g. archiving, backup, display, power supply or detection functions (e.g. of sabotage, opening). For this purpose, the security degree of the alarm system is the result of the assessment of three core areas, namely the electronic security system and the emergency alarm system (see Table 2), the video surveillance system (see Table 3) and the electronic access control system (see Table 4).

The security factor of the Security Forces is their degree of preparedness and equipment and armament (L_{SF}). Each of the security forces has certain competences in relation to the response to intruders based on the legislation in the given country. At the same time, depending on the operator, the financial possibilities that are invested in motivation, education, practical training, certification, equipment and armament of members of the security forces differ. It follows from the above that the degree of powers, preparedness and equipment increases from the provision of own protection by the critical infrastructure operator (Level 1), through private security services (Level 2), to the provision of protection by state security forces, e.g. municipal police, state by the police (Level 3) or the military (Level 4).

The degree of their standardization can be a security factor of Regime Measures (L_{RM}). It can be regime measures introduced on the basis of informal, or customary rules (Level 1) or it can be formalized processes that have their own owner (e.g.

Table 3 Security level of the video surveillance system (VVS) [45]

Security level	Risk level	Requirements	Object type
Level 1	Low risk	The VSS system has no intrusion protection, and monitoring of essential functions is not required	Small storage areas (<400 m^2) for products of low attractiveness (vegetables or newspapers). Enterprises whose activities do not indicate an impact on values (life, health, property) or confidential information (e.g. sugar factory)
Level 2	Low to medium risk	The VSS system has simple intrusion protection and no monitoring of basic functions is required	Large storage spaces (>400 m^2) for products of low attractiveness. Businesses whose activities do not indicate an impact on values or confidential information (e.g. medical laboratories, paper mills, collecting raw materials, etc.)
Level 3	Medium to high risk	The VSS system has moderate protection against intrusion and monitoring of basic functions is required	Large storage areas (>400 m^2) for products of medium attractiveness (shopping centre). Businesses whose activities may affect values but not confidential information (e.g. public buildings, ports, airports, banks, etc.)
Level 4	High risk	The VSS system has high protection against intrusion and monitoring of basic functions is required	Storage spaces for highly attractive products (jewellery, regulated medicines). Businesses whose activities may affect values and confidential information (e.g. mobile phone sales point, military laboratories, government buildings, etc.)

Chief Security Officer, Chief Information Security Officer) and form (e.g. established structure of security documentation—strategies, policies, directives, regulations, manuals, projects, news) (Level 2). In the case of the introduction of formalized processes related to regime measures, it is possible to subject them to an independent audit according to the relevant national certification scheme (Level 3) or an international certification scheme, e.g. ISO/IEC 27,001 [51] (Level 4).

Table 4 Electronic access control system (EACS) security level [48]

Security level	Risk level	Application	Intruder ability	Object type
Level 1	Low risk	Organizational measures, protection of low value assets	Poor knowledge of ACS, identification means and IT technologies	Hotel
Level 2	Low to medium risk	Organizational measures, protection of low to medium value assets	Small funds to carry out the attack	Business offices, small businesses
Level 3	Medium to high risk	Fewer organizational measures, protection of medium to high value commercial assets	Intermediate knowledge of ACS, identification means and IT technologies	Industry, administrative premises, financial institutions
Level 4	High risk	Protection of very high value commercial assets or critical infrastructure	Small to medium funds to implement the attack	Highly sensitive areas (military installations, government buildings, research and development, critical infrastructure)

A security factor for detection capability may be the Cumulative Probability of DETection (P_{CPDET}). After overcoming the protected perimeter to reaching the protected interest, the intruder often passes through several detection zones with different detection probabilities P_{Di}. In such a case, we can talk about the cumulative probability of correct intruder detection, which represents the overall probability of his detection before reaching his goal [41]. This quantity can be calculated according to Eq. (1):

$$P_{CPDET} = \left[1 - \prod_{i=1}^{n}(1 - P_{Di})\right] \cdot P_{FS} \cdot P_{FT} \cdot P_{HAR} \tag{1}$$

where P_{CPDET} = Cumulative Probability of (correct) intruder DETection [0-1]; n = number of detection zones during the intruder's path [number]; P_{Di} = probability of correct Detection by an active element (e.g. passive infrared sensor) in the ith detection zone during the intruder's path [0-1]; P_{FS} = probability of a Fault-free State of the alarm system [0-1]; P_{FT} = probability of Fault-free Transmission of the alarm signal via the alarm transmission path to the remote alarm receiving centre [0-1];

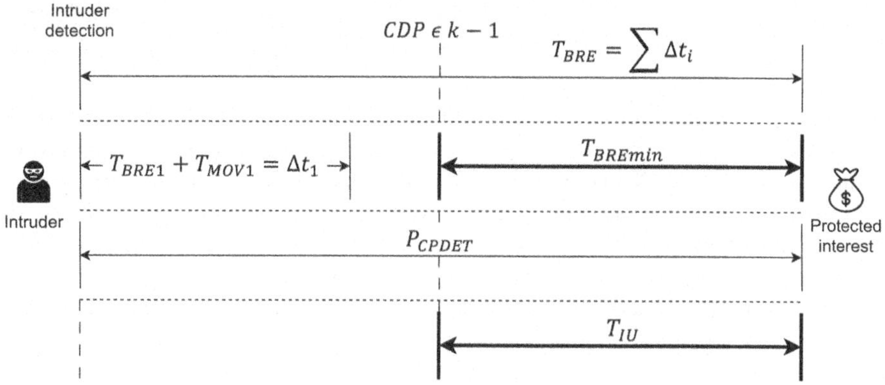

Fig. 3 Graphic representation of the critical detection point [41]. *Legend* T_{BRE1} = the first barrier BREaking time from the moment of intruder detection [s]; T_{MOV1} = the intruder MOVement time after breaking the first barrier [s]; Δt_i = ith barrier breaking time or intruder movement times on the way to the protected interest which meet the condition that their sum is just greater than the response time of the intervention unit [s]; T_{BREmin} = minimum total time to BREak through the mechanical protection barriers [s]; P_{CPDET} = cumulative probability of (correct) intruder detection [0-1]; T_{IU} = total response time of the Intervention Unit [s]; n = total number of passive protection elements [number]; k = protection passive element before which an intruder must be detected in order for the system to still be functional (from the point of view of the breakthrough resistance of the system and the reaction time of the intervention unit)

P_{HAR} = probability of the Human Agent Reliability [0-1]. Note that (1) describes that detection of the intruder in any pathway segment suffices. However, it is assumed that the transmission of the alarm signal is functioning throughout as well as its communication system.

Another security factor of detection capability is the Critical Detection Point time (T_{CDP}). The CDP point (see Fig. 3) expresses the physical protection system property that, if the system is to be functional, detection must occur at the latest at this point/element or at previous active elements [25].

The position of the active element $k - 1$ is determined by fulfilling the condition that the intruder's delay time when moving and overcoming mechanical barriers (T_{BREmin}), from this point to the protected interest (point n), just exceeded the response time of the response unit (T_{IU}) [41]. From a design point of view, the worst case scenario is when the critical detection point is in the immediate vicinity of the protected interest. On the contrary, the ideal case is when the critical detection point is located even before the object perimeter.

The security factor of the capability response is the probability of eliminating the intruder ($P_{R|A}$). This parameter defines how likely an intruder is to be intercepted or otherwise eliminated during his journey to the protected interest. For the success of the intervention unit, it follows that a random variable X must be greater than 0 [25]. Then this probability can be determined according to Eq. (2), when assuming a normal distribution of X:

$$P_{R|A} = P(X > 0) = \int_{0}^{\infty} \frac{1}{\sqrt{2\pi\sigma_x^2}} \cdot e^{\frac{-(x-\mu_x)^2}{2\sigma_x^2}} dx \tag{2}$$

where $P_{R|A}$ = the probability of a successful response of the intervention unit [0-1]; μ = mean value of a random variable X; σ^2 = variance (dispersion, variation) of a random variable X. Probability $P_{R|A}$ is based on the basic evaluation criterion of the functionality of the physical protection system, i.e. that the time of the attack, or possibly the total time to break through internal and external barriers T_{BRE}, must be greater than the reaction time of the intervention unit T_{IU} [25].

The conditions stated in relation (3) follows from the mentioned criterion and has been used when defining (2):

$$T_{BRE} - T_{IU} > 0 \tag{3}$$

where T_{BRE} = the total time of the attack by the intruder from the time of detection by alarm systems, until he leaves the guarded area [s]; T_{IU} = total response time of the emergency unit [s]. Times T_{BRE} and T_{IU} are independent random variables that have a normal distribution with parameters μ, σ^2. . This creates a new random variable X, which has the same distribution as the given times. For the new random variable X the following relationships apply (4)–(6):

$$X = T_{BRE} - T_{IU} \tag{4}$$

$$\mu_x = E(T_{BRE} - T_{IU}) = E(T_{BRE}) - E(T_{IU}) \tag{5}$$

$$\sigma_x^2 = D(T_{BRE} - T_{IU}) = D(T_{BRE}) - D(T_{IU}) \tag{6}$$

where μ = mean value of a random variable X; σ^2 = variance (dispersion, variation) of a random variable X; T_{BRE} = the total time of the attack, including the time to overcome the barriers and move the intruder [s]; T_{IU} = total response time of the emergency unit [s].

5 Assessing the Critical Infrastructures Elements Resilience Using Security Factors of the Physical Protection System

Assessing the critical infrastructures elements resilience can currently be implemented using a number of supporting methods. These methods are mainly focused on assessing static resilience [52]. This means that they provide information about the resilience level that the element achieves in the time before the disruptive event. The most suitable methods can be considered to include: Resilience Assessment in Electricity Critical Infrastructure from the Point of View of Converged Security [53]; A Performance-based Tabular Approach for Joint Systematic Improvement of Risk Control and Resilience Applied to Telecommunication Grid, Gas Network, and Ultrasound Localization System [54]; Assessing and Strengthening Organisational Resilience–ASOR Method [55]; Critical Infrastructure Elements Resilience Assessment—CIERA Method [18]; Availability-based engineering resilience metric and its corresponding evaluation methodology [56]; Resilience Capacities Assessment for Critical Infrastructures Disruption: The READ Framework [57]; A Quantitative Method for Assessing Resilience of Interdependent Infrastructures [58]; Guidelines for Critical Infrastructure Resilience Evaluation [59]; Measuring Critical Infrastructure Resilience: Possible Indicators [60]; and Resilience Measurement Index—RMI [61].

In the context of the solved issue, i.e. the use of the physical protection system to assess the critical infrastructures elements resilience, the authors would like to follow up on their previous research. For this reason, the CIERA method [18] was chosen, which enables a semi-quantitative assessment of the elements resilience in the context of specific disruptive events. Using the findings from this method, the authors in the following part of the text define (1) the assessment parameters of the security factors of the physical protection system, and (2) the procedure for their use in assessing the critical infrastructures elements resilience against intentional threats of a physical nature, e.g. terrorist attack.

5.1 Assessment Parameters of the Physical Protection System Security Factors

Due to the diversity of the identified security factors, their assessment parameters are also quite different. For this reason, it was first necessary to set up a uniform assessment scale that will enable subsequent semi-quantitative assessment of the critical infrastructures elements resilience using the physical protection system. The essence of this scale is a linear five-point scale, where the value 1 represents the lowest possible resilience level of the security factor and the value 5, on the contrary, the highest possible resilience level of the security factor. Defining the assessment parameters of the physical protection system security factors enabling the assessment of the resilience level of these factors is presented in Table 5.

Based on the definition of standardized assessment parameters, it is possible to assess the resilience level of physical protection system's individual security factors of the critical infrastructures elements. Using these values, it is possible to further determine the resilience levels of individual components and their variables. For this purpose, however, it is first necessary to define the procedure and mathematical relationships for assessing the level of static resilience of critical infrastructure elements.

Table 5 Assessment parameters of the physical protection system security factors

Security factors of the physical protection system	Assessment parameters of the security factors resilience level
Breakthrough resistance time of mechanical barriers (T_{MB})	5: Delay time of the intruder on path to the protected interest T_{DEL} reaches significantly higher values than the reaction time of the intervention unit T_{REA}
	4: Delay time of the intruder on path to the protected interest T_{DEL} reaches higher values than the reaction time of the intervention unit T_{REA}
	3: Delay time of the intruder on path to the protected interest T_{DEL} it reaches the same value as the reaction time of the intervention unit T_{REA}
	2: Delay time of the intruder on path to the protected interest T_{DEL} reaches slightly lower values than the reaction time of the intervention unit T_{REA}
	1: Delay time of the intruder on path to the protected interest T_{DEL} reaches significantly lower values than the reaction time of the intervention unit T_{REA}
Security level of the alarm system (L_{AS})	5: The security of all three alarm systems (see Tables 2, 3 and 4) reaches Level 4, which means that they are able to withstand a high risk of being breached
	4: The lowest achieved level of security for alarm systems (see Tables 2, 3 and 4) is Level 3, which means that the least secure alarm system is able to withstand a medium to high risk of intrusion
	3: The lowest achieved level of security for alarm systems (see Tables 2, 3 and 4) is Level 2, which means that the least secure alarm system is able to withstand a low to medium risk of intrusion
	2: The lowest achieved level of security for alarm systems (see Tables 2, 3 and 4) is Level 1, which means that the least secure alarm system is able to withstand only a low risk of intrusion
	1: The critical infrastructure element has no security alarm system installed, meaning it is unable to withstand even a low risk of intrusion
Preparedness and equipment level of the security forces (L_{SF})	5: Physical protection of the critical infrastructure element is provided by the military security forces of the state
	4: Physical protection of the critical infrastructure element is provided by the police security forces of the state (state or municipal police)

(continued)

Table 5 (continued)

Security factors of the physical protection system	Assessment parameters of the security factors resilience level
	3: Physical protection of the critical infrastructure element is provided by private security services
	2: The physical protection of the critical infrastructure element is ensured by the own security components of the entity (owner/operator) of the critical infrastructure
	1: The physical protection of the critical infrastructure element is not provided by any security components
Level of regime measures standardization (L_{RM})	5: Regime measures have been implemented based on formalized processes and have been independently audited according to an international certification scheme (e.g. ISO/IEC 27,001)
	4: Regime measures have been implemented based on formalized processes and have been independently audited according to the relevant national certification scheme
	3: Regime measures were introduced based on formalized processes
	2: Regime measures were introduced based on informal/customary rules
	1: No regulatory measures have been implemented for the critical infrastructure element
Cumulative probability of detection (P_{KDET})	5: The cumulative probability of intruder detection reaches 96–100%
	4: The cumulative probability of intruder detection reaches 91–95%
	3: The cumulative probability of intruder detection reaches 81–90%
	2: Cumulative probability of intruder detection reaches 51–80%
	1: Cumulative probability of intruder detection reaches 0–50%
Critical detection point (T_{CDP})	5: The critical point of intruder detection is located in front of the object's perimeter
	4: The critical point of intruder detection is located on the perimeter of the object
	3: The critical point of intruder detection is located in the external or internal areas of the protected interest

(continued)

Table 5 (continued)

Security factors of the physical protection system	Assessment parameters of the security factors resilience level
	2: The critical point of intruder detection is located in the immediate vicinity of the protected interest
	1: The critical point of intruder detection is identical to the protected point of interest
Probability of the intruder elimination ($P_{R\|A}$)	5: Probability of the intruder elimination reaches 81–100%
	4: Probability of the intruder elimination reaches 61–80%
	3: Probability of the intruder elimination reaches 41–60%
	2: Probability of the intruder elimination intruder is 21–40%
	1: Probability of the intruder elimination intruder reaches 0–20%

5.2 *The Procedure for Assessing the Critical Infrastructures Elements Resilience Against Intentional Threats of a Physical Nature*

As already stated at the beginning of this part of the chapter, the mechanism for assessing the critical infrastructure elements resilience against threats of a physical nature will be based primarily on the CIERA method [18]. The benefit of this condition is the synchronization of assessment mechanisms, which enables the backward integration of newly defined security factors among the existing measurable items of the CIERA method. This synchronization thus enables not only the use of the assessment mechanism for assessing the critical infrastructures elements resilience in the context of the physical protection system, but also the retroactive extension of the CIERA method by the resistance component, which is not included in this method.

The essence of the procedure for assessing the critical infrastructure elements resilience against physical threats is the gradual aggregation of individual resilience factors. In the present case 7 security factors, 4 variables and 2 components (see Table 1) into one data, which is the resulting level of critical infrastructure element resilience against physical threats.

The first stage of the procedure is to determine the critical infrastructure element resilience level variables introduced in Table 1 which are calculated as a weighted average of the security factors concerned (Relation 7). Since the level of security factors is expressed by a point rating, a percentage transformation must be performed in this calculation to convert the five-point scale into percentage points, i.e. multiplying the resulting point value by 20.

Table 6 Normalized weights of physical protection system security factors

Critical infrastructure element resilience variables (V_j)	Physical protection system security factors (SF_k)	Normalized weights (w_k)
Physical resistance (V_{PR})	Breakthrough resistance time of mechanical barriers (T_{MB})	1.0
Security measures (V_{SM})	Security level of the alarm system (L_{AS})	0.4
	Preparedness and equipment level of the security forces (L_{SF})	0.4
	Level of regime measures standardization (L_{RM})	0.2
Detection capability (V_{DC})	Cumulative probability of detection (P_{KDET})	0.4
	Critical detection point (T_{CDP})	0.6
Responsiveness (V_{RN})	Probability of the intruder elimination ($P_{R\vert A}$)	1.0

$$V_j = 20 \cdot \sum_{k=1}^{l} SF_k \cdot w_k \tag{7}$$

where V_j = jth variable of the critical infrastructure element resilience [%]; SF_k = kth Security Factor of the critical infrastructure element resilience [number of points] according to Table 5; w_k = kth normalized weight of the kth security factor of the critical infrastructure element resilience [0-1]; l = the total number of security factors in the jth variable.

The normalized weights of security factors determine their relative importance in determining the critical infrastructure element resilience variables. The general rule for determining these weights is that their sum in the variable is equal to 1. These weights were determined using the pairwise comparison method [62] and their values for individual security factors are presented in Table 6.

A possible graphical representation of the resulting levels of critical infrastructure element resilience variables is presented in Fig. 4.

The second stage of the procedure is determining the level of the components, i.e. resistance and robustness introduced in Table 1. The levels of individual critical infrastructure element resilience components are determined by the weighted average of their variables (Eq. 8):

$$C_i = \sum_{j=1}^{m} V_j \cdot w_j \tag{8}$$

where C_i = the ith component of critical infrastructure element resilience [%]; V_j = jth variable critical infrastructure element resilience [%]; w_j = jth normalized weight

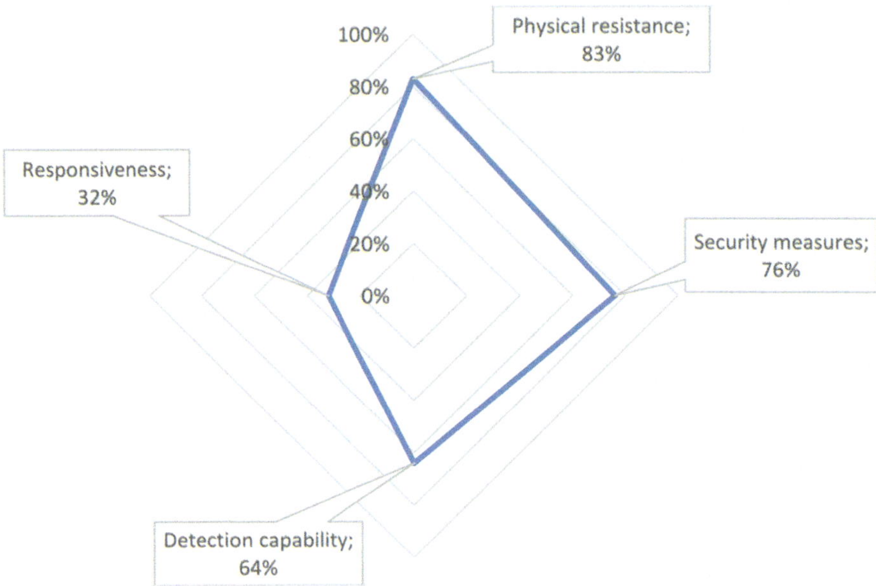

Fig. 4 Graphic representation of critical infrastructure element resilience variables level

of the jth critical infrastructure element resilience variable [0-1]; m = the total number of variables for the ith component.

The normalized weights of variables determine their relative importance in determining the critical infrastructure element resilience components. The general rule for determining these weights is that their sum in the variable is equal to 1. These weights were also determined using the pairwise comparison method [62] and their values for individual security factors are presented in Table 7.

The third stage of the procedure is the determination of the critical infrastructure element resilience level, which can be considered as a composite indicator of the assessed critical infrastructure elements' physical protection level. The level of critical infrastructure element resilience is calculated as the arithmetic mean of the values of the components by which it is determined (Eq. 9):

Table 7 Normalized weights of critical infrastructure element resilience variables

Critical infrastructure element resilience components (C_i)	Critical infrastructure element resilience variables (V_j)	Normalized weights (w_j)
Resistance (C_{RES})	Physical resistance (V_{PR})	0.4
	Security measures (V_{SM})	0.6
Robustness (C_{ROB})	Detection capability (V_{DC})	0.4
	Responsiveness (V_{RN})	0.6

Table 8 Comparison table for assessing the critical infrastructure element resilience level [18]

The critical infrastructure element resilience level (%)	Assessing the resilience level
85–100	High level of resilience
69–84	An acceptable level of resilience
53–68	Low level of resilience
37–52	Insufficient level of resilience
≤36	Critical level of resilience

$$R = \frac{1}{n} \cdot \sum_{i=1}^{n} C_i \tag{9}$$

where $R =$ critical infrastructure element resilience [%]; $C_i =$ the ith component of critical infrastructure element resilience [%]; $n =$ the total number of components determining resilience.

The final stage of the procedure is the assessment of the achieved resilience level, indicating the critical infrastructure element physical protection level. To assess the resilience level, it is possible to use the comparison table from the CIERA method (see Table 8).

In conclusion, it must be stated that if the resilience reaches a level of $\leq 52\%$, it is necessary to identify weak points consisting in the decomposition of the resilience assessment results, which must be carried out at the security factors level concerned. For security factors showing a resilience level with a point value of 2 or less, it is necessary to revise them and start the process of their strengthening [63] or total restoration.

6 Conclusion

The successful operation of the Smart City concept is dependent on a number of external factors from its very beginning. Energy and ICT infrastructure can be considered as the most important of them in the field of technical support. For this reason, elements of these infrastructures must have a high level of resilience to the effects of current security threats, especially intentional threats of a physical nature. For this purpose, the concept of assessing the critical infrastructure elements resilience by assessing their physical protection system is presented in this chapter. The essence of this concept is the definition of the physical protection system security factors and the procedure for their use in assessing the critical infrastructures elements resilience against intentional physical threats, e.g. a terrorist attack.

The physical protection system security factors defined in this chapter were divided into four areas, which are physical resistance, security measures, detection capability, and response capability. A total of seven security factors were identified and specified in these areas, which provide basic information on the level of critical infrastructure elements resilience. These factors are (i) breakthrough resistance time of mechanical barriers, (ii) security level of the alarm system, (iii) preparedness and equipment level of the security forces, (iv) level of regime measures standardization, (v) cumulative probability of detection, (vi) critical detection point, and (vii) probability of the intruder elimination. Assessment parameters were subsequently determined for these factors and they were classified into a linear five-level scale for the purpose of semi-quantitative assessment of the critical infrastructures elements resilience. Also, the factors where weighted within each area. The total resilience of a critical infrastructure is obtained by determining these factors for each infrastructure element (critical entities of similar importance) and computing an average.

The concept of assessing the critical infrastructure elements resilience using the physical protection system is an assessment tool designed for critical infrastructure entities security managers, e.g. Security Liaison Officer. Due to its semi-quantitative and focused nature it is an expert-assessment based feasible, slim and efficient approach. This tool is used to assess primary risks disrupting the system of critical infrastructures elements physical protection. Based on the output information, it is possible to identify the weak points of this system and start the process of strengthening the critical infrastructure elements resilience.

Future research should be focused in particular on the identification and specification of other security factors suitable for assessing the critical infrastructures elements resilience within or beyond the presented framework and research on tools suitable for strengthening resilience. Future work could also concentrate on approaches beyond semi-quantification or how to use quantitative results within semi-quantitative approaches.

Acknowledgements This work was supported by the Ministry of the Interior of the Czech Republic [grant number VI20192022151].

References

1. Neuman M (2020) Infrastructure is key to make cities sustainable. Sustainability 12:8308. https://doi.org/10.3390/su12208308
2. Greed C (2022) The importance of social infrastructure in supporting physical infrastructure and enabling the development of sustainable cities. In: Iossifova D, Gasparatos A, Zavos S, Gamal Y, Long Y (eds) Urban infrastructuring. sustainable development goals series, Springer, Singapore, pp 265–276. https://doi.org/10.1007/978-981-16-8352-7_16
3. Häring I, Schäfer J, Vogelbacher G, Fischer K, Riedel W, Faist K, Fehling-Kaschek M, Scharte B, Ross K, Stolz A, Leismann T, Hiermaier S (2021) From event to performance function-based resilience analysis and improvement processes for more sustainable systems. Int J Sustain Mater Struct Syst 5(1–2):90–120. https://doi.org/10.1504/IJSMSS.2021.115784
4. Council of the European Union (2008) Council Directive 2008/114/EC of 8 December 2008 on the identification and designation of European critical infrastructures and the assessment of the need to improve their protection

5. Ersoy A (2017) Smart cities as a mechanism towards a broader understanding of infrastructure interdependencies. Reg Stud Reg Sci 4(1):26–31. https://doi.org/10.1080/21681376.2017.1281154
6. Czupich M (2019) The role of ICT in the smart city concept. Olsztyn Econ J 14(1):63–74. https://doi.org/10.31648/oej.3646
7. Tomforde S, Gelhausen P, Gruhl C, Häring I, Sick B (2019) Explicit consideration of organic computing in design process. In: 32nd international conference on architecture of computing systems, pp 51–56. https://ieeexplore.ieee.org/abstract/document/8836197
8. Zalitis Z, Dolgicers A, Zemite L, Ganter S, Kopustinskas V, Vamanu B, Finger J, Fuggini C, Bode I, Kozadajevs J, Häring I (2022) Mitigation of the impact of disturbances in gas transmission systems. Int J Crit Infrastruct Prot 39:100569. https://doi.org/10.1016/j.ijcip.2022.100569
9. Ward DM (2013) The effect of weather on grid systems and the reliability of electricity supply. Clim Change 121(1):103–113. https://doi.org/10.1007/s10584-013-0916-z
10. Morgan MG et al (2012) Terrorism and the electric power delivery system. National Academy of Sciences, Washington, DC. https://doi.org/10.17226/12050
11. Siebold U, Ziehm J, Häring I (2009) Terror event database and analysis software. In Elsner, P. (ed) 4th future security research conference, pp 85–93
12. Häring I (2015) Introduction to database analysis of terroristic events. In: Risk analysis and management: engineering resilience, Springer, Singapore, pp 27–60. https://doi.org/10.1007/978-981-10-0015-7_3
13. Jain AK, Grumber C, Gelhausen P, Häring I, Stolz A (2020) A toy model study for long-term terror event time series prediction with CNN. Eur J Secur Res 5:289–309. https://doi.org/10.1007/s41125-019-00061-w
14. Kampova K, Lovecek T, Rehak D (2020) Quantitative approach to physical protection systems assessment of critical infrastructure elements: use case in the slovak republic. Int J Crit Infrastruct Prot 30:100376. https://doi.org/10.1016/j.ijcip.2020.100376
15. Holling CS (1973) Resilience and stability of ecological systems. Annu Rev Ecol Syst 4:1–23. https://doi.org/10.1146/annurev.es.04.110173.000245
16. Häring I, Ebenhöch S, Stolz A (2016) Quantification of resilience for resilience engineering of socio technical systems. Eur J Secur Res 1(1):21–58. https://doi.org/10.1007/s41125-015-0001-x
17. NIAC (2009) Critical infrastructure resilience final report and recommendations. National Infrastructure Advisory Council, Washington, DC.
18. Rehak D, Senovsky P, Hromada M, Lovecek T (2019) Complex approach to assessing resilience of critical infrastructure elements. Int J Crit Infrastruct Prot 25:125–138. https://doi.org/10.1016/j.ijcip.2019.03.003
19. Häring I, Schäffer L, Restayn EM, Vogelbacher G, Stolz A, Finger J (2021) Formalization of questionnaire-based score card risk control and resilience assessment for critical infrastructure operators and companies countering covid-19. In: Castanier B, Cepin M, Bigaud D, Berenguerpp C (eds) 31th European safety and reliability conference (ESREL 2021), pp 2862–2869. https://doi.org/10.3850/978-981-18-2016-8_314-cd
20. ISO 31000 (2018) Risk management–guidelines. International Organization for Standardization, Geneva
21. Häring I (2021) Technical safety, reliability and resilience: methods and processes. Springer, Singapore. https://link.springer.com/book, https://doi.org/10.1007/978-981-33-4272-9
22. Rehak D, Senovsky P, Slivkova S (2018) Resilience of critical infrastructure elements and its main factors. Systems 6(2):21. https://doi.org/10.3390/systems6020021
23. Aguilar FJ (1967) Scanning the business environment. Macmillan, New York, NY
24. Bugdol M, Jedynak P (2015) Integrated management systems. Springer, Cham. https://doi.org/10.1007/978-3-319-10028-9
25. Garcia ML (2008) The design and evaluation of physical protection systems, 2nd ed. Elsevier, Oxford. https://doi.org/10.1016/C2009-0-25612-1

26. Commission of the European Communities (2005) Green Paper on a European programme for critical infrastructure protection (COM/2005/0576 final).
27. Government of the Slovak Republic (2008) The concept of critical infrastructure in the Slovak Republic and ways of its protection and defence.
28. Nirandjan S, Koks EE, Ward PJ, Aerts JCJH (2022) A spatially-explicit harmonized global dataset of critical infrastructure. Sci Data 9:150. https://doi.org/10.1038/s41597-022-01218-4
29. Batista e Silva F, Forzieri G, Marin Herrera MA, Bianchi A, Lavalle C, Feyen L (2019) HARCI-EU, a harmonized gridded dataset of critical infrastructures in Europe for large-scale risk assessments. Sci Data 6:126. https://doi.org/10.1038/s41597-019-0135-1
30. Bennett HA (1977) EASI: approach to physical security evaluation. [Technical Report SAND-76-0500]. Sandia Labs, Albuquerque, NM
31. Matter C (1988) SAVI: A PC-based vulnerability assessment program. [Technical Report SAND-88-1279]. Sandia Labs, Albuquerque, NM
32. Winblad AE (1989) SAVI: systematic analysis of vulnerability to intrusion. [Technical Report SAND-89-0926]. Sandia Labs, Albuquerque, NM
33. Al-Ayat RA, Cousins TC, Matter JC (1989) An overview of ASSESS-analytic system and software for evaluating safeguards and security. [Technical Report UCRL-101256]. Lawrence Livermore National Laboratory, Livermore, CA
34. Al-Ayat RA, Cousins TD, Hoover ER (1990) ASSESS update–current status and future developments. [Technical Report UCRL-JC-104360]. Lawrence Livermore National Laboratory, Livermore, CA
35. Tarasov Y (2008) Specialized software systems. Security, reliability, information 78(3)
36. Jang SS, Kwan SW, Yoo H, Kim JS, Yoon WK (2009) Development of a vulnerability assessment code for a physical protection system: systematic analysis of physical protection effectiveness (SAPE). Nucl Eng Technol 41(5):747–752. https://doi.org/10.5516/NET.2009.41.5.747
37. Vintr Z, Valis D, Malach J (2012) Attack tree-based evaluation of physical protection systems vulnerability. In: IEEE international carnahan conference on security technology (ICCST), pp 59–65. https://doi.org/10.1109/CCST.2012.6393538
38. Lovecek T, Ristvej J, Simak L (2010) Critical infrastructure protection systems effectiveness evaluation. J Homel Secur Emerg Manag 7(1):1–25. https://doi.org/10.2202/1547-7355.1613
39. Lovecek T, Siser A, Maris L (2017) Use case of waterwork physical protection system robustness evaluation as a part of slovak critical infrastructure. In: 51st IEEE international carnahan conference on security technology (ICCST), Madrid, Spain, pp 1–5. https://doi.org/10.1109/CCST.2017.8167811
40. Vintr Z, Machal J (2012) Evaluation of physical protection system effectiveness. In: IEEE international carnahan conference on security technology (ICCST), pp 16–21. https://doi.org/10.1109/CCST.2012.6393532
41. Lovecek T, Reitspis J (2011) Designing and evaluation of object protection systems. University of Zilina, Zilina
42. Vlk I, Martin M (2007) Guide to the protection of objects. Digital Graphic, Trencin
43. EN 50136-1 (2012) Alarm systems–alarm transmission systems and equipment–part 1: general requirements for alarm transmission systems. European Committee for Electrotechnical Standardization, Brussels
44. EN 50398-1 (2017) Alarm systems–combined and integrated alarm systems–part 1: general requirements. European Committee for Electrotechnical Standardization, Brussels
45. EN 62676-1-1 (2014) Video surveillance systems for use in security applications–part 1-1: system requirements–general. European Committee for Electrotechnical Standardization, Brussels
46. EN 50130-4 (2011) Alarm systems–part 4: electromagnetic compatibility–product family standard: Immunity requirements for components of fire, intruder, hold up, CCTV, access control and social alarm systems. European Committee for Electrotechnical Standardization, Brussels
47. EN 50131-1 (2006) Alarm systems. Intrusion systems–part 1: system requirements. European Committee for Electrotechnical Standardization, Brussels

48. EN 60839-11-1 (2013) Alarm and electronic security systems–part 11-1: electronic access control systems–system and components requirements European Committee for Electrotechnical Standardization, Brussels
49. EN 50134-1 (2002) Alarm systems–social alarm systems–part 1: system requirements European Committee for Electrotechnical Standardization, Brussels
50. Selinger P, Uhrin S (2003) Security services. EDIS, Zilina.
51. ISO/IEC 27001 (2022) Information security, cybersecurity and privacy protection–information security management systems–requirements. International Organization for Standardization, Geneva
52. Simonovic SP, Arunkumar R (2016) Comparison of static and dynamic resilience for a multi-purpose reservoir operation. Water Resour Res 52(11):8630–8649. https://doi.org/10.1002/2016WR019551
53. Hromada M, Rehak D, Lukas L (2021) Resilience assessment in electricity critical infrastructure from the point of view of converged security. Energies 14:1624. https://doi.org/10.3390/en14061624
54. Haring I, Fehling-Kaschek M, Miller N et al (2021) A Performance-based tabular approach for joint systematic improvement of risk control and resilience applied to telecommunication grid, gas network, and ultrasound localization system. Environ Syst Decis 41:286–329. https://doi.org/10.1007/s10669-021-09811-5
55. Rehak D (2020) Assessing and strengthening organisational resilience in a critical infrastructure system: case study of the Slovak Republic. Saf Sci 123:104573. https://doi.org/10.1016/j.ssci.2019.104573
56. Cai B, Xie M, Liu Y, Liu Y, Feng Q (2018) Availability-based engineering resilience metric and its corresponding evaluation methodology. Reliab Eng Syst Saf 172:216–224. https://doi.org/10.1016/j.ress.2017.12.021
57. Kozine I, Petrenj B, Trucco P (2018) Resilience capacities assessment for critical infrastructures disruption: the READ framework. Int J Crit Infrastruct 14(3):199–220. https://doi.org/10.1504/IJCIS.2018.10015604
58. Nan C, Sansavini G (2017) A quantitative method for assessing resilience of interdependent infrastructures. Reliab Eng Syst Saf 157:35–53. https://doi.org/10.1016/j.ress.2016.08.013
59. Bertocchi G, Bologna S, Carducci G, Carrozzi L, Cavallini S, Lazari A, Oliva G, Traballesi A (2016) Guidelines for critical infrastructure resilience evaluation. Italian Association of Critical Infrastructures' Experts, Roma
60. Prior T (2015) Measuring critical infrastructure resilience: possible indicators (Risk and Resilience Report 9). Eidgenössische Technische Hochschule, Zurich
61. Petit F, Bassett G, Black R, Buehring W, Collins M, Dickinson D, Fisher R, Haffenden R, Huttenga A, Klett M, Phillips J, Thomas M, Veselka S, Wallace K, Whitfield R, Peerenboom J (2013) Resilience measurement index: an indicator of critical infrastructure resilience. Argonne National Laboratory, Lemont, IL
62. Saaty TL (1980) The analytic hierarchy process: planning, priority setting, resource allocation. McGraw-Hill, New York, NY
63. Rehak D, Slivkova S, Janeckova H, Stuberova D, Hromada M (2022) Strengthening resilience in the energy critical infrastructure: methodological overview. Energies 15(14):5276. https://doi.org/10.3390/en15145276

David Rehak is a Ph.D. in Processes Modelling and Simulation of Troops and Population Protection at the University of Defence. He is Full Professor at the VSB–Technical University of Ostrava, Faculty of Safety Engineering. His scientific and research work is aimed on critical infrastructure resilience & protection, risk management, civil protection and environmental protection. He is for example member of the International Association of Critical Infrastructure Protection Professionals (IACIPP), member of the Czech Technology Platform Energy Security (TPEB), editor-in-chief of the Transactions of the VSB-Technical University of Ostrava, Safety Engineering Series (TSES), reviewer of the International Journal of Critical Infrastructure Protection (IJCIP), reviewer of the Journal of Infrastructure Systems (JIS) and reviewer of the International Journal of Disaster Risk Reduction (IJDRR).

Tomas Lovecek is a Ph.D. in Quality Evaluation of Physical Protection Systems at the University of Žilina. He is Full Professor at University of Žilina, Faculty of Security Engineering. His scientific and research work is aimed on physical protection systems, risk management, alarm system, modelling and simulation. He was project coordinator of the EACEA project Competency Based e-portal of Security and Safety Engineering and partner coordinator of two projects funded by EC/DG Home Affairs, PACITA: Methodology for physical protection assessment of critical infrastructure elements against terrorist and other types of attacks and CI-PAC: Critical Infrastructure Protection Against Chemical Attack. He is author of scientific monographs "Designing and Evaluation of the Physical Protection Systems" and "Physical Protection of Critical Transport Infrastructure". He is a member of editorial boards for scientific journals: European Journal of Security and Safety, Communications, Journal of Criminal Justice and Security and Crisis Management.

Martin Hromada graduated at Tomas Bata University in Zlín in study field Security technologies, systems and management. He defended the dissertation thesis "Technological Aspects of Critical Infrastructure Protection of the SR" in 2011. He worked as a consultant of Deloitte Advisory and currently works as a lecturer at the Department of Security Engineering, Faculty of Applied Informatics, Tomas Bata University in Zlín. In 2017 he defended his habilitation work at the Faculty of Safety Engineering, VSB–Technical University of Ostrava and received an academic title of Associate Professor in the field of Safety and fire protection. Within the framework of research activities, he actively focuses on the protection and resilience of critical (information) infrastructure and the evaluation of functionality of physical protection systems.

Neil Walker has over 30 years experience in events and strategic marketing communications, event management, media and publishing, with the past 15 years being focused in the security and aerospace sectors. Founder and CEO of Torch Marketing Co Ltd, an integrated b2b strategic marketing communications agency, Neil is also founder and director of the World Border Security Congress and Border Security Report magazine, as well as founder and director of the World Border Security Network and International Association of Critical Infrastructure Protection Professionals. A long history of driving forward successful events in various sectors from international sporting events to energy, aerospace and security sectors, Neil continues to deliver community developing projects and programmes in the border security and critical infrastructure protection communities.

Ivo Häring holds a Ph.D. of the Max Planck Institute for the Physics of Complex Systems (MPIKS) and the Technical University Dresden (TUD). In 2004 he started as research scientist and then project lead at Fraunhofer Institute for High-Speed Dynamics, Ernst-Mach-Institute, EMI. From 2008 till 2013 he lead a research group on Technical Safety and from 2015 till 2019 on Resilience Engineering. Since 2014 he is Senior Scientist, currently within the Department of Safety and Resilience. Being heavily involved in EU and national civil security research he was involved in the development and execution of such EU projects as VITRUV, BESECURE, XP-DITE, D-BOX, ENCOUNTER, EDEN, SnowBall, AF3, RESILIENS, TRESSPASS, RESISTO, SecureGas and eFORT and the German national projects SUSQRA, STADTSICHERHEIT 3D and OCTIKT. Results were presented in numerous (conference) publications and also in lectures at the Department of Sustainable Systems Engineering (INATECH) of the University of Freiburg and within the Risk Engineering master course at the University of Applied science Furtwangen mainly in the domains of functional safety, quantitative and probabilistic risk and resilience analysis.

Smart City Security Based on a Meta-Security Framework for Digital Twins

Keizo Sugiyama, Masataka Nakahara, Norihiro Okui, Yasuaki Kobayashi, and Ayumu Kubota

Abstract Security measures in the context of smart cities have been taken mainly in cyberspace; these measures include alerting users based on cyber threat information and observed events on the network. As the integration of cyberspace and physical space progresses towards beyond 5th generation (B5G) era, attacks and their effects will expand not only to cyberspace but also to physical space, and it is expected that completely new security threats will emerge. In this paper, we introduce a meta-security framework that spans cyberspace and physical space as a novel security measure for smart cities using digital twins. This framework is intended to offer comprehensive security solutions to oversee and harmonize multiple security mechanisms in both spaces.

Keywords Digital twin · Meta-security · Smart city

List of Acronyms

3D	Three-dimensional
4G	4th generation
5G	5th generation

K. Sugiyama (✉) · M. Nakahara · N. Okui · Y. Kobayashi · A. Kubota
KDDI Research, Inc. 2-1-15 Ohara, Fujimino-Shi, Saitama 356-8502, Japan
e-mail: sugiyama@kddi-research.jp

M. Nakahara
e-mail: ms-nakahara@kddi-research.jp

N. Okui
e-mail: no-okui@kddi-research.jp

Y. Kobayashi
e-mail: yas-kobayashi@kddi-research.jp

A. Kubota
e-mail: kubota@kddi-research.jp

R. Shinkuma et al. (eds.), *Advances in Engineering and Information Science Toward Smart City and Beyond*, Engineering Cyber-Physical Systems and Critical Infrastructures 5, https://doi.org/10.1007/978-3-031-29301-6_2

AI	Artificial Intelligence
B5G	Beyond 5th Generation
C&C	Command and Control
CPS	Cyber-Physical System
DDoS	Distributed DoS
DoS	Denial of Service
DTN	Digital Twin Network
EC	Electronic Commerce
ETSI	European Telecommunications Standards Institute
GAN	Generative Adversarial Network
GDPR	General Data Protection Regulation
ICT	Information and Communication Technology
IIC	Industrial Internet Consortium
IoT	Internet of Things
IP	Internet Protocol
ITU-T	International Telecommunication Union Telecommunication Standardization Sector
LAN	Local Area Network
LiDAR	Light Detection And Ranging
MANO	Management and Orchestration
MIC	Ministry of Internal Affairs and Communications
NICT	National Institute of Information and Communications Technology
NIST	National Institute of Standards and Technology
NOTICE	National Operation Toward IoT Clean Environment
NPN	Nonpublic Network
OS	Operating System
PCF	Policy Control Function
PLM	Product Lifecycle Management
SMF	Session Management Function
VIM	Virtualized Infrastructure Manager
VNF	Virtual Network Function
VNFM	VNF Manager
XR	Cross reality

Glossary of Most Important Terms

Security for digital twins	Securing digital twins
Digital twins for security	Utilizing digital twins as security measures
Meta-security	A technology to control and harmonize multiple independent security mechanisms in physical space and cyberspace

1 Introduction

The development of information and communication technology (ICT) is very rapid and has made remarkable progress in recent decades. As a result, our lives today depend on many services provided in cyberspace. Various services in cyberspace are penetrating our daily lives; these services include electronic commerce (EC); the distribution of content, such as video, music, and games; and the ability to instantly communicate with people far away via the internet.

It is expected that the future direction of ICT is to establish social and economic infrastructure by integrating cyberspace and physical space. Society 5.0 [1], Japan's proposed vision of society in the 2030s, aims for an inclusive, sustainable, and dependable society based on a cyber-physical system (CPS). Internet of Things (IoT) devices are used in CPSs to convert events and phenomena in physical space into data, which are stored and analyzed in cyberspace to enhance services and values and then fed back into physical space. For example, a user wears a smart device on his or her wrist. Pulse, acceleration, location data, etc., from the smart device are accumulated in cyberspace for a certain period and analyzed. Based on the analysis results, information on his or her health condition can be provided to the user, who lives in physical space. It is also possible to provide his or her family with feedback and guidance. Smart cities are the means of expanding such efforts to regions and cities to lead to the resolution of diverse social issues, such as rising medical costs and aging populations, the revitalization of communities, mobility, disaster prevention, and the promotion of decarbonization. In Japan, the government is promoting smart city initiatives as a comprehensive showcase for Society 5.0.

Security is one of the essential technologies and must be a fundamental part of a CPS. The National Institute of Standards and Technology (NIST) in the U.S. mentions that security is a necessary feature of the CPS architecture to ensure that CPS capabilities are not compromised by malicious agents and that the information used, processed, stored, and transferred has its integrity preserved and kept confidential where needed [2]. To date, cybersecurity measures have been taken from various perspectives, including attack detection and anomaly detection over wide area networks.

The increase in the number of IoT devices will make it necessary to take countermeasures from many more perspectives. Towards B5G era, which is expected to be introduced around 2030, various IoT devices, such as sensors and actuators, are connected to the network through B5G, and cyberspace and physical space are more highly integrated. Attacks and their impacts will expand not only in cyberspace but also in physical space, and completely new security threats that have never occurred before are expected to emerge. It will be difficult to recognize such threats and take countermeasures only with information observable in cyberspace, and a security mechanism that coordinates cyberspace and physical space will be necessary. For example, KDDI Corporation and KDDI Research, Inc. have proposed a next-generation social concept called KDDI Accelerate 5.0 [3]. Seven technologies (namely, network, security, IoT, platform, artificial intelligence (AI), cross reality

(XR), and robotics) are orchestrated to optimize social and economic activities by circulating information and data between cyberspace and physical space. In the future, in addition to ensuring the security of cyberspace, it will be necessary to ensure the security of physical space, and to implement security measures considering the interaction between cyberspace and physical space. Next-generation security technologies are required to protect people and society from more sophisticated attacks.

To address the above issues, this paper proposes the concept of meta-security to oversee and harmonize multiple security mechanisms. A meta-security framework extends across physical security layer and cybersecurity layer and interacts with physical applications and digital twin applications. A meta-security component manages whether security mechanisms in both layers are working properly and, if not, takes security measures according to the situation.

The remainder of this paper is organized as follows. Section 2 introduces some research and development cases related to smart cities and digital twins and classify the model of digital twins. Section 3 states the problem of the existing digital twin security. Then, we propose the concept of meta-security for digital twins in Sect. 4. In addition, a case study of security threats across cyberspace and physical space is presented for personal mobility as an example of the application of meta-security. In Sect. 5, the future direction of smart city security is explained. Finally, Sect. 6 concludes the paper.

2 Review of Smart City Security and Digital Twins

The development of smart cities is gaining momentum in cities around the world. One possible disadvantage of the realization of smart cities, however, is the concern that the collection of all kinds of information may lead to an invasion of privacy. As a secure social infrastructure, ensuring security is indispensable to protect privacy properly.

2.1 Smart City Security

The Smart City Security Guideline [4] was published by the Ministry of Internal Affairs and Communications (MIC), Japan. The guideline explains security concepts and security measures in smart city promotion to realize safe and secure smart cities and describes security not only in technology aspects but also in management aspects by relying on the Smart City Reference Architecture [5]. As an approach to examine the security of smart cities, the hierarchy defined in Smart City Reference Architecture is organized into four categories, i.e., governance, service, city operating system (OS) and assets from a security perspective, as shown in Fig. 1. Security is defined as an element that relates to all other items and requires detailed consideration.

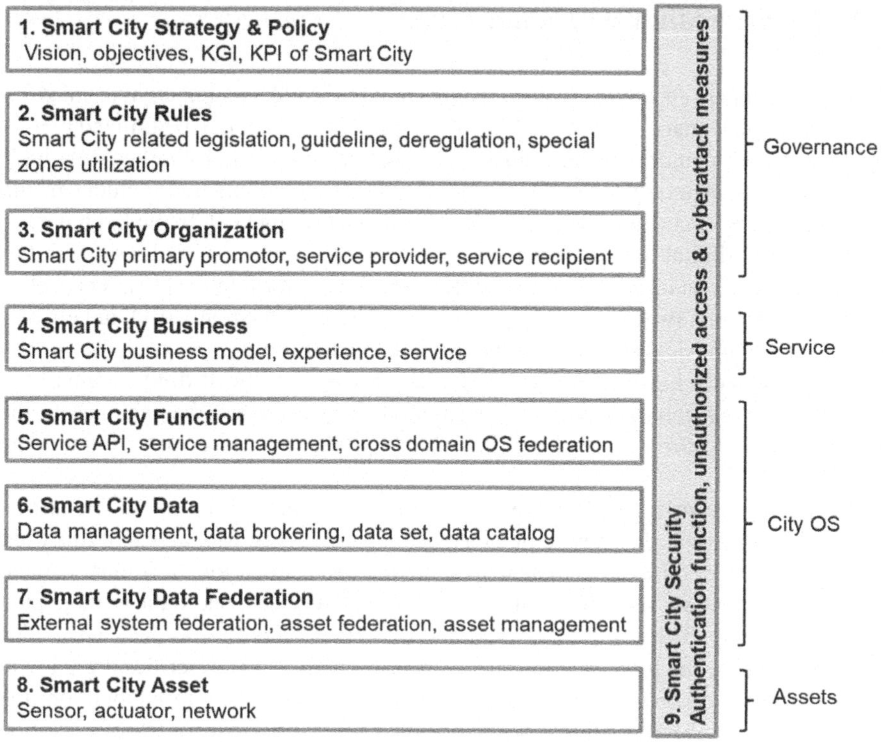

Fig. 1 Items to be defined in the Smart City Reference Architecture and four categories of security

Standardization activities on digital twins in smart cities and networks are also conducted by the International Telecommunication Union Telecommunication Standardization Sector (ITU-T). It is appropriate to make use of digital twins in studies on smart cities since smart cities are one of the distinctive applications of CPSs and digital twins are one of the core technologies driving CPSs. Security guidelines for digital twin networks (DTNs) have been discussed; these guidelines describe security considerations, requirements, and measures for networks of digital twins [6]. As DTNs will also face security risks and various attacks and it is required to be secure enough to avoid all possible attacks, these guidelines categorize threats into four groups: threats to layers, threats to interfaces, threats to digitalization and threats to management.

Several research papers have been published on smart city security. State-of-the-art security solutions for smart city applications such as privacy-preserving infection spread analysis are introduced [7]. The smart cities' challenges in areas relating to privacy, security, and risk are analyzed from various stakeholder perspectives [8]. A platform architecture which integrates a number of security solutions and respects general data protection regulation (GDPR) is presented [9].

2.2 Security Study on Digital Twins

The basic concept of digital twins was presented in 2002 as a conceptual ideal for product lifecycle management (PLM) [10]. Digital twins began to be utilized mainly by users in the manufacturing industry [11] but are now being adopted in a wide range of fields. The target domain of digital twins is expanding and includes health care [12], energy management [13] and autonomous driving [14]. In the Digital Twin Consortium, a digital twin is defined as a virtual representation of real-world entities and processes, synchronized at a specified frequency and fidelity [15]. The IoT is used to construct a twin in cyberspace that is paired with the real world and enables monitoring in physical space and simulation in cyberspace.

Although there have been several studies on the security of digital twins, it is still an emerging field, and further contributions are necessary. We can classify the activities into "security for digital twins" and "digital twins for security" to facilitate the review.

Regarding "security for digital twins", a survey on potential security threats associated with digital twins has been conducted [16]. In the survey, four layers of functionality have been identified, i.e., data dissemination and acquisition, data management and synchronization, data modeling and additional services, and data visualization and accessibility. Threats are analyzed and classified according to each layer. The journal issued by the Industrial Internet Consortium (IIC) [17] describes digital twin threats and steps to securing a digital twin implementation. The journal mentions the risk in which the hacker can spoof the behavior of digital twins or even access the physical system, thereby potentially providing access to system-wide data.

Regarding "digital twins for security", digital twins are used as security measures. A preliminary conceptual framework was introduced to increase the level of security by using a cyber digital twin providing advanced anomaly detection capabilities by using state-of-the-art machine learning solutions [18]. A generative adversarial network (GAN) based approach is introduced to improve anomaly detection with digital twins [19]. Another study applied a digital twin for industrial networks to provide a playground for cyber-security analysis and validation, based on the European Telecommunications Standards Institute (ETSI) Network Function Virtualization (NFV)—Management and Orchestration (MANO) approach [20]. The usability of digital twins for cybersecurity and their potential for cybersecurity incident prediction is also reported in [21]. There are few approaches to using digital twins for security measures.

2.3 Models of Digital Twins

The ways to construct digital twins vary depending on the services and the applications. Numerous digital twin architectures have been proposed thus far. As one attempt to classify them, a three-dimensional digital twin reference architecture

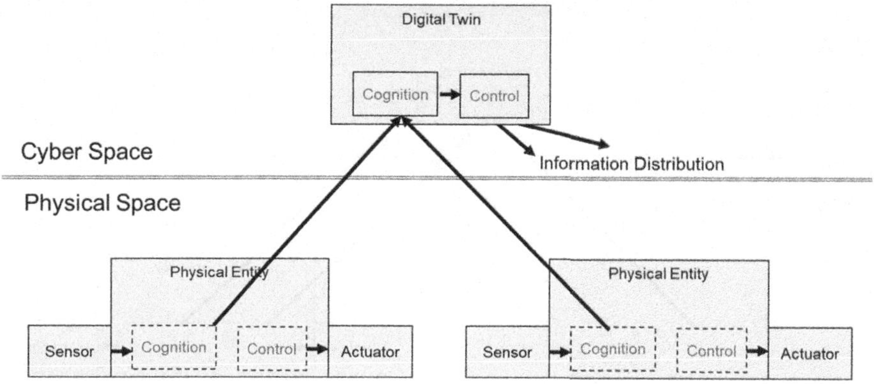

Fig. 2 Pseudo digital twin

model for applications across industries and that considers functionality, dependability, and life cycle aspects was proposed [22]. The purpose of this model is to allow practitioners to plan, develop, and implement digital twin applications, independent of the field, the use case, or the complexity of the application.

Digital twins allow cyberspace and physical space to interact with each other. Therefore, the essential elements to construct digital twins are sensors for sensing the state of physical space continuously or intermittently, networks for bridging physical space and cyberspace, digital twins in cyberspace and actuators for controlling entities in physical space. A digital twin in cyberspace includes a cognition function for analyzing data from sensors and a control function for supplying actuators with instructions, probably with the help of AI.

Here, existing digital twin models can be classified into pseudo digital twins, general digital twins, and proper digital twins, in terms of the interaction between physical space and cyberspace, as shown in Figs. 2, 3 and 4. This classification contributes to the study of how digital twins affect physical space and cyberspace.

- **Pseudo digital twin**

Data in physical space are obtained by sensors in each physical entity. The obtained data are formatted to make them easier to handle in cyberspace and to transmit. A digital twin entity in cyberspace collects data through networks and conducts analysis by using the data. Pseudo digital twins can simulate situations that are difficult to verify in physical space, but feedback based on the analysis results is closed within cyberspace and is provided only to the other entities within cyberspace. This model is a kind of cloud service since there is no direct impact on entities in physical space. Typical examples are location-based coupon distribution and itinerary suggestions for smartphones and digital signage.

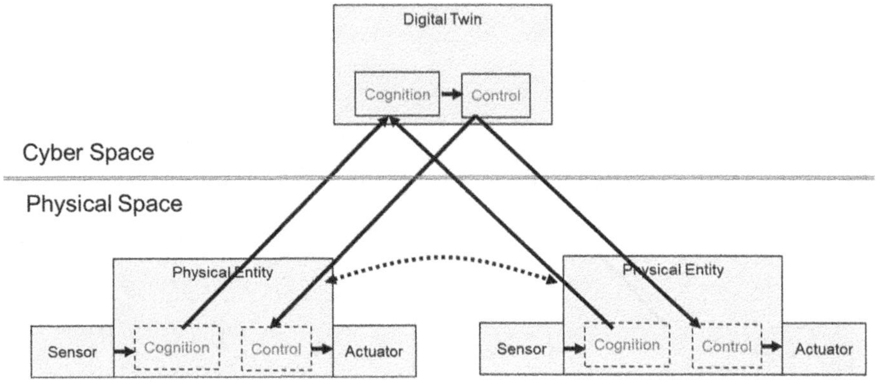

Fig. 3 General digital twin

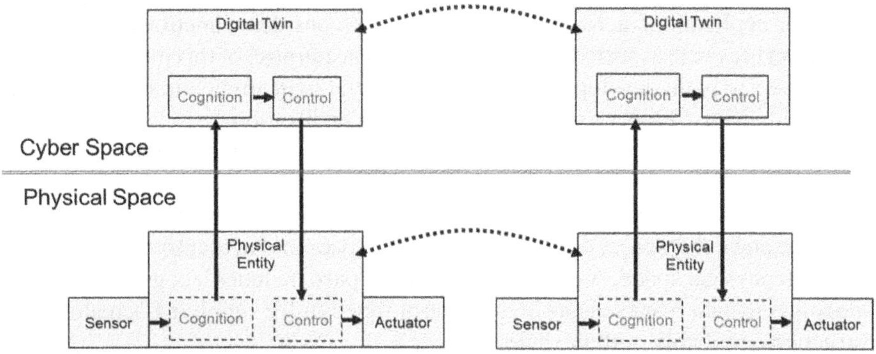

Fig. 4 Proper digital twin

- **General digital twin**

The first half of the general digital twin process is the same as the pseudo digital twin process. The difference is that after the obtained data are collected in a digital twin of the cyberspace and analyzed, feedback of the analysis results is transmitted to the physical space to control the physical entities. In other words, feedback influences the entity in physical space. There may be indirect interactions among physical entities via a digital twin (as indicated by the dotted arrow in Fig. 3). For example, a camera mounted on a drone can act as a sensor to capture images of conditions at a construction site. These images are analyzed in a digital twin entity to ensure safety. On the basis of the results of that analysis, construction equipment in physical space can be remotely operated via the digital twin entity. In addition, the drone can be automatically steered according to the operation of the construction equipment.

- **Proper digital twin**

Unlike general digital twins, digital twin entities in cyberspace are constructed to mirror each physical entity. There are interactions among the digital twin entities in cyberspace as a matter of course (as indicated by the dotted arrow in Fig. 4). Their interactions can be extended to handle new functionalities that are not mere replicas of activities in physical entities. Fine-grained feedback is given to each physical entity and influences the physical space. There may be direct or indirect interactions among physical entities. By utilizing a proper digital twin, it is possible to comprehensively represent complex systems from various aspects and to predict possible future events through repeated interactions.

3 Problem Statement

Conventionally, security for cyberspace and physical space has been considered independently. As mentioned in 2.2, there have been several studies on digital twin security, but they are mainly focused on the single space or one-directional impacts, such as the impacts from cyberspace to physical space. In such cases, the following problems exist:

- Even if individual security measures are taken in cyberspace and physical space, they may be insufficient in some cases; i.e., security measures should be taken in both spaces while considering bidirectional influence, while a single security measure in each space may be sufficient in other cases. An example of the former case is that if an IoT device has a rogue circuit installed and is communicating with the command and control (C&C) server, it is necessary both to take down the C&C server and to deactivate the rogue circuit. An example of the latter is a model inversion attack on an AI model in cyberspace; this type of attack can usually be handled only in cyberspace.
- It is necessary to manage whether security mechanisms existing in both cyberspace and physical space are working properly. If a system detects that a security mechanism in one space is not working properly, the system may be able to shut down the mechanism and allow other mechanisms to compensate for the defective mechanism. For example, if the behavior of a device management application in cyberspace has become abnormal due to being infected with some malware, the application will be stopped and the security mechanism in physical space will activate auxiliary measures by using another device.
- Prioritization of countermeasures against threats that occur in cyberspace and physical space may be required in some cases. For example, when serious cyber-attacks on power grids in smart cities occur and traffic signals stop, restoration by a backup power supply should be conducted first to prevent accidents from occurring, and security measures should be taken next. Harmonization among multiple

security mechanisms in both spaces should be necessary since cyberspace and physical space interact with each other.

Considering the above, we propose the concept of meta-security in Sect. 4, a technology to control and harmonize multiple independent security mechanisms in physical space and cyberspace.

4 Introduction of Meta-Security for Digital Twins

This section first explains a model of meta-security in the context of digital twins. Although conventional security measures for cyberspace and physical space have been considered independently, a meta-security framework extends across physical security layer and cybersecurity layer. A meta-security component is introduced to coordinate and take security measures across the layers.

As digital twins are often adopted as systems that support the activities in smart cities, the model can be applied to most smart city applications. An example that embodies a meta-security model is next provided to facilitate the understanding of the concept.

4.1 Model of Meta-Security

In physical space, sensors and/or actuators are driven by physical applications. Security mechanisms may be included in physical applications; for example, a mechanism that performs integrity checks may be used in IoT device firmware, i.e., sensors and actuators, to protect against tampering. Data acquired in physical space are transmitted to digital twin applications by the network from physical space to cyberspace and, in turn, are received by physical applications to drive actuators. In cyberspace, digital twin applications reflect the status of physical entities and drive virtual sensors and virtual actuators. In Fig. 5, virtual sensors and virtual actuators correspond to the cognition function and control function, respectively. Security mechanisms, such as authentication and access control, must be implemented in digital twin applications. There may be differences in terminology, but thus far, this configuration does not differ from the usual digital twin configuration.

The newly introduced component in this model is the meta-security box on the right side in Fig. 5. This component extends across the physical security layer and cybersecurity layer and interacts with physical applications and digital twin applications to manage whether security mechanisms existing in both layers are working properly and, if not, take security measures according to the situation. To do so, the meta-security component has physical security managers and cybersecurity managers who monitor the correct operations of each of these security mechanisms in both layers. Furthermore, a meta-security manager mediates between

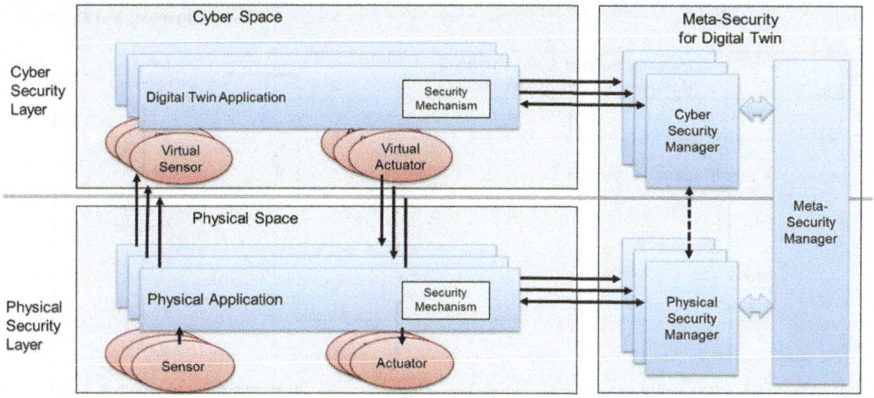

Fig. 5 Model of meta-security for digital twins

the physical security managers and the cybersecurity managers to ensure security, conducting prioritization of security measures against threats in physical space and/or cyberspace.

As the technical feature of the proposed model is to make use of a meta-security component for coordinated activities, along with the physical applications and the digital twin applications, this model will be applicable to most smart city applications and act as a complementary method for existing security measures. The main focus of this model is to achieve "security for digital twins".

A similar concept can be found in the telecom industries as shown in Fig. 6 [23]. ETSI adopts NFV-MANO for mobile core networks to manage and orchestrate various virtual network functions (VNFs), such as the session management function (SMF) and policy control function (PCF). NFV-MANO allows us to deploy various network services with different requirements and characteristics and to achieve operational sophistication and advanced network control. MANO consists of the virtualized infrastructure manager (VIM), VNF Manager (VNFM), and NFV Orchestrator and plays a role in collecting mobile network data, analyzing the entire network, and making configuration change for network optimization. This approach enables one to manage and control multiple functions and layers from a network management point of view.

4.2 Use Cases of Meta-Security Framework in Smart City

Micromobility can be taken as a typical example of a smart city use case because several demonstration tests are being conducted to verify the safety and social acceptability of micromobility [24]. Mobility scooters are single-seat electric vehicles with three or four wheels and are often used as alternative transportation for personal mobility. Mobility scooters support autonomous driving functions and are suitable

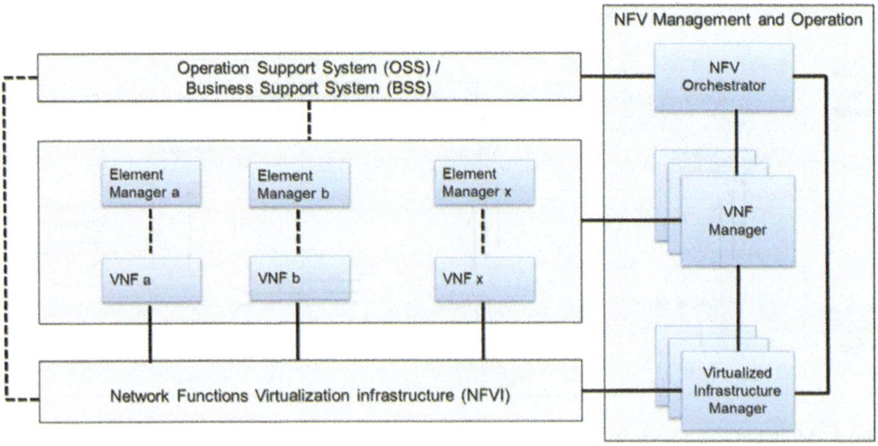

Fig. 6 NFV architecture framework

for people, especially seniors, who want to get around quickly, easily, and cheaply while avoiding congestion on urban streets in large shopping malls, during hospital visits, during neighborhood excursions, and during other city-wide activities.

The global market for micromobility is growing, and the need for autonomous driving technology is increasing. However, as an increasing number of vehicles are mixed in a certain area, there is concern that a high level of security and safety is needed, and the pace of diffusion will slow down. B5G is expected to be useful for connected cars and automated driving in terms of simultaneous multiple connectivity and low latency, but their security in the context of B5G is still in the discussion stage and will require further research and development.

In this section, we examine how the abovementioned meta-security model can be applied to the mobility scooter's use case and increase the level of security.

In physical space, as shown in Fig. 7, a mobility scooter is equipped with sensors such as light detection and ranging (LiDAR) sensors, and cameras to identify self-location in collaboration with a digital twin application in cyberspace. The digital twin application is driven by interaction with physical sensors. The digital twin application collects the sensor data from the mobility scooter and estimates the relative locations of other mobility scooters, and feeds the instructions back to the target mobility scooter. Examples of threats to these sensors are jamming signals, deterioration over time, and malware infections of firmware; and examples of threats to digital twin applications are malware infection and processing delays due to network congestion. These threats may cause data falsification. The physical security manager and the cybersecurity manager for the mobility scooter monitor the operation of the security mechanisms at each layer and inform the status of the meta-security manager. The meta-security manager comprehensively secures multiple use cases, not just specific layers or use cases.

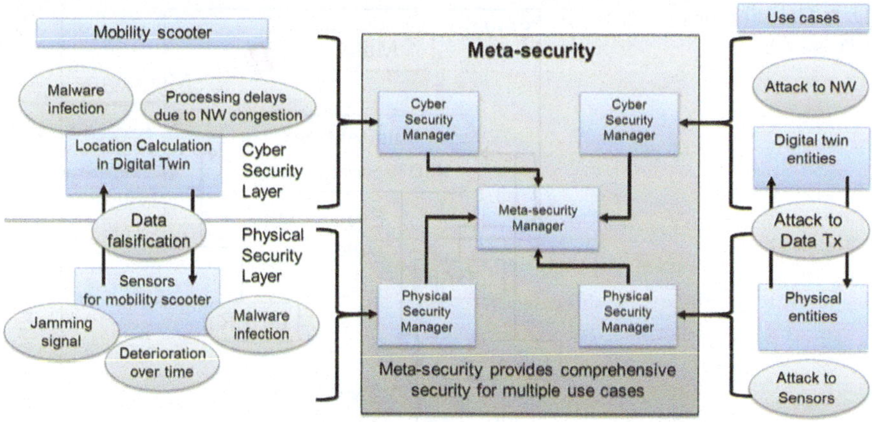

Fig. 7 Application of meta-security to actual use cases

Figure 8 shows a detailed explanation of how meta-security works. Normally, LiDAR measures the exact distance to an object by reflecting laser light and converts the light into a three-dimensional (3D) point measurement. LiDAR measurement data are collected in a server. The server processes 3D reconstruction from them and transmits the information to mobility scooters. The feedback from the digital twin is delayed and augmented by meta-security. The following paragraph describes the case of processing delay and/or data transmission delay.

- **Precondition**

In physical space, a LiDAR sensor on a mobility scooter experiences some kind of malfunction due to jamming or degradation. Measurement results are no longer available in some of the directions that the scooter's LiDAR is responsible for. Otherwise, in cyberspace, computing resources in a server running the digital twin application have been exhausted. This causes processing delays to 3D reconstruction from LiDAR data. The server may be infected with malware or under a denial of service (DoS) attack.

- **Observed phenomena**

The arrival delay of the 3D reconstruction data to the mobility scooter caused the route to be incorrectly calculated. The deviation of the mobility scooter's travel path has increased. Collisions with other mobility scooters may occur.

- **Coordination by the meta-security mechanism**

The physical security manager receives the information from the mobility scooter that some LiDAR sensors are in a state of malfunction. The cybersecurity manager receives no response or information from the digital twin application that the computing resource is full. This information is abstracted and reported to the meta-security manager. Considering the situation comprehensively, the probability of an

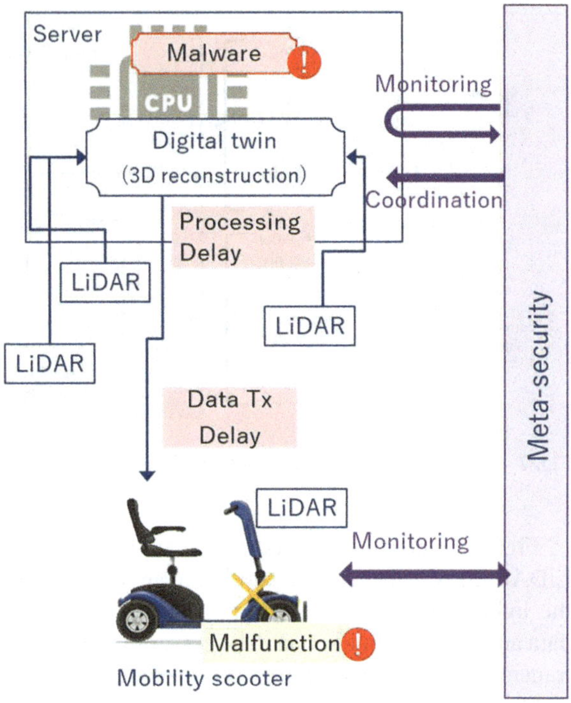

Fig. 8 Coordination by meta-security

accident (e.g., collision or overstepping) is estimated; consequently, one or more of the following decisions are made according to urgency and importance:

1. To transmit instructions to lead to a stop or to stop transmitting data to each security manager (in the latter case, its fail-safe mechanism is expected to stop the mobility scooter)
2. To increase the advance width of the prediction calculation in the digital twin application (i.e., increase the number of seconds ahead to generate LiDAR reconstruction data)
3. To determine the type of malware causing the high load from cyber-threat intelligence information and notify the cybersecurity manager of countermeasures.

Finally, instructions are applied to the mobility scooter and the digital twin application via each security manager to avoid potential risks.

5 Future Direction of Smart City Security

This section describes how digital twins can be applied to smart cities as "digital twins for security" to ensure security in a society where cyberspace and physical space are highly integrated by B5G.

Current IoT security efforts focus mainly on technologies that can be handled in wide-area cyberspace. For example, vulnerable IoT devices are discovered from the internet by a port scan at a certain internet protocol (IP) address range, and abnormal communications from malware-infected IoT devices are detected on the internet as distributed DoS (DDoS) attacks. The Ministry of Internal Affairs and Communications (MIC) and the National Institute of Information and Communications Technology (NICT) in Japan have conducted the "NOTICE" (National Operation Toward IoT Clean Environment) project to survey vulnerable IoT devices and to alert users to the problem [25].

Insufficient consideration, however, has been given to the behavior of IoT devices that can be observed only in physical space and near IoT devices, such as within local area networks (LANs) and nonpublic networks (NPNs). Although attacks are not observed from wide-area networks, such cases are expected to increase gradually in B5G, where cyberspace and physical space will converge. The following attacks and measures are possible:

- Attacks are made on IoT devices in physical space without being observed from wide-area cyber networks. For example, although an anomaly of an IoT device is only detected within a local network in some facilities, similar devices in another site may have the same situation and cause incidents before long. To prevent damage from incidents, nearby detection information is useful for sharing with other sites and deployed over a wide area.
- Even if a problem is not detected from the cyber perspective, there may be a problem from the physical perspective. For example, if an unauthorized circuit is embedded in the design of a device and a backdoor is installed to be penetrated from the outside, it is difficult to detect the problem from the cyber perspective alone. Security can be improved by deploying the threats found from the physical perspective, such as rogue circuit detection, over a wide area.

There have been no activities to monitor and take countermeasures against the above attacks and their large impacts. Even if an attack or unauthorized behavior is observed locally, it is difficult to accurately determine whether there is an impact on the wide area network and whether countermeasures are necessary. Just as micromobility will be equipped with a variety of sensors and actuators, it is expected that the number of combinations of devices that can create problems and threats will increase. However, it will become more difficult to respond to all possible threats as the integration between cyberspace and physical space progresses. It will be necessary to realize a new security infrastructure that deploys threat information detected in physical space and locally in cyberspace over a wide area and uses this information to detect and counter potential threats by other similar devices that cannot directly detect threats. It is also necessary to prioritize countermeasures based on the size and frequency of threats.

To solve this problem, we can construct a digital twin using the information obtained in both cyberspace and physical space. Information about nearby devices can be reflected in the digital twin in a form that can be used for security measures, and threats can be analyzed, including information obtained at other locations and

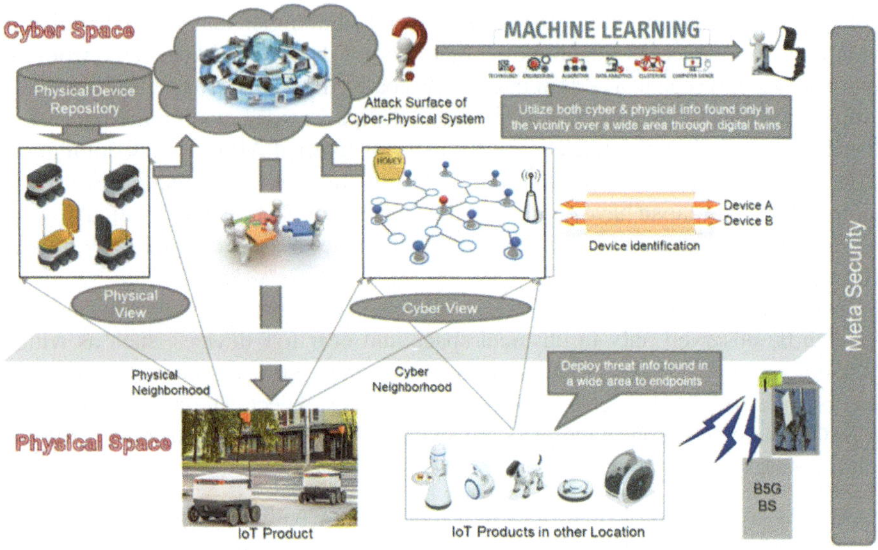

Fig. 9 Digital twin for smart city security

in a wide area, to enable rapid response to threats. Figure 9 shows an image of the proposed platform.

IoT devices are regarded not only as IP communication nodes but also as entities in physical space, including the information on onboard firmware and circuit information; i.e., IoT devices are regarded from not only a cyber view but also a physical view. The digital twin analyzes the relationship and impact between security threats discovered elsewhere and the device in question. From the profile information of individual IoT devices and the observation information of cyberspace and physical space, we generate digital twins necessary for analyzing the relationship and impact between security threats discovered elsewhere and the relevant devices. Digital twins can be used for security analysis and countermeasures for IoT systems and device groups in other locations.

In this scheme, meta-security plays an important role in protecting against attacks in a smart city where multiple types of IoT devices coexist. We assume that delivery robots are used to deliver food to tenant companies in office buildings, as shown in Fig. 9. There exist security threats that specifically target robotic systems [26]. Although the roads may be partially used, these delivery robots are not subject to motor vehicle regulations. Therefore, they could potentially collide with cars due to sensor malfunctioning. To reduce such risks, a digital twin is constructed for each IoT use case. As depicted in Fig. 7, meta-security provides comprehensive security for multiple use cases. The digital twin of delivery robots and the digital twin of traffic flow control coordinate their information via a meta-security component to facilitate and secure the movement of the delivery robots.

The goal of this infrastructure is to construct a platform for security measures by using the digital twin method, which maps and represents information about IoT devices in physical space onto cyberspace, and to enhance security measures for CPSs in society. Digital twins can be used to respond quickly to new threats and incidents to prevent or minimize damage.

6 Conclusion

Secure development and secure lifecycle management are being enhanced for IoT devices, and the robustness of IoT devices themselves is gradually increasing. However, in the B5G era, compared to the current 5G era, more devices are expected to be connected to the network. It is highly likely that critical and diverse devices used in social infrastructure will be connected to the network. Therefore, robustness will not be able to keep up with the increase in the number and diversity of devices, and many vulnerable IoT devices will still exist.

To realize a safe and secure society in smart cities, it is essential to ensure not only information security in cyberspace but also comprehensive security in both physical space and cyberspace. However, there are still various gaps between their security domains, and many issues must be addressed when considering physical security and cybersecurity. In this paper, a meta-security framework is proposed to coordinate cybersecurity and physical security in each of these domains. This framework will be applicable to most smart city applications and act as a complementary method for the existing security measures. The use of digital twins to secure smart cities is indicated as a future direction toward an infrastructure for autonomous security measures.

Acknowledgements We acknowledge Professor Ryoichi Shinkuma of Shibaura Institute of Technology and Associate Professor Takayuki Nishio of Tokyo Institute of Technology for their suggestions and constructive discussion. These research results were partly obtained from the commissioned research (No.05201) by the National Institute of Information and Communications Technology (NICT), Japan.

References

1. Society 5.0. https://www8.cao.go.jp/cstp/english/society5_0/index.html
2. NIST Special Publication 1500-201 (2017) Framework for cyber-physical systems, vol 1, Overview, U.S. Department of Commerce
3. Formulation of "KDDI Accelerate 5.0," with the aim of creating a resilient future society in the "New Normal" Era. https://news.kddi.com/kddi/corporate/english/newsrelease/2020/08/31/4658.html
4. Smart City Security Guideline (Ver 1.0), MIC, Japan, 2020. https://www.soumu.go.jp/main_sosiki/joho_tsusin/eng/presentation/pdf/Smart_City_Security_Guideline_ver1.0.pdf

5. Smart City Reference Architecture White Paper (First edition), CAO, Japan, 2020. https://www8.cao.go.jp/cstp/stmain/a-whitepaper1_200331_en.pdf
6. Security Guidelines for Digital Twin Network (2021) draft new recommendation ITU-T X.sg-dtn TD 4057.
7. Zhang K et al (2017) Security and privacy in smart city applications: challenges and solutions. IEEE Commun Mag 55(1):122–129. https://doi.org/10.1109/MCOM.2017.1600267CM
8. Ismagilova E et al (2022) Security, privacy and risks within smart cities: literature review and development of a smart city interaction framework. Inf Syst Front 24:393–414. https://doi.org/10.1007/s10796-020-10044-1
9. Badii C et al (2020) Smart city IoT XE "IoT" platform respecting GDPR privacy and security aspects. IEEE Access 8:23601–23623. https://doi.org/10.1109/ACCESS.2020.2968741
10. Grieves M, Vickers J (2017) Digital twin: mitigating unpredictable, undesirable emergent behavior in complex systems. In: Trans-disciplinary perspectives on complex systems. Springer, Berlin, Germany
11. Qi Q et al (2018) Digital twin service towards smart manufacturing. In: 51st CIRP conference on manufacturing systems
12. Erol T et al (2020) The digital twin revolution in healthcare. In: 4th international symposium on multidisciplinary studies and innovative technologies (ISMSIT)
13. Yu W et al (2022) Energy digital twin technology for industrial energy management: classification, challenges and future. In: Renew Sustain Energy Rev 161(C). Elsevier
14. Veledar O et al (2019) Digital twins for dependability improvement of autonomous driving. In: European conference on software process improvement, pp 415–426
15. "Definition of a Digital Twin", Digital Twin Consortium. https://www.digitaltwinconsortium.org/initiatives/the-definition-of-a-digital-twin/
16. Alcaraz C, Lopez J (2022) Digital twin: a comprehensive survey of security threats. IEEE Commun Surv Tutor 24(3)
17. Hearn M, Rix S (2019) Cybersecurity considerations for digital twin implementations. IIC J Innov
18. Pokhrel A et al (2020) Digital twin for cybersecurity incident prediction: a multivocal literature review. In: IEEE/ACM 42nd international conference on software engineering workshops
19. Xu Q et al (2021) Digital twin-based anomaly detection in cyber-physical systems. In: 2021 14th IEEE conference on software testing, verification and validation (ICST). IEEE
20. Benedictis A et al (2022) Toward the adoption of secure cyber digital twins to enhance cyber-physical systems security. In: International conference on the quality of information and communications technology (QUATIC)
21. Grasselli C et al (2022) An industrial network digital twin for enhanced security of cyber-physical systems. In: International symposium on networks, computers and communications (ISNCC)
22. Newrzella SR et al (2017) Three-dimension digital twin reference architecture model for functionality, dependability, and life cycle development across industries. IEEE Access
23. ETSI GS NFV-MAN 001 (2014) Network Functions Virtualisation (NFV); Management and Orchestration V1.1.1 2014-12
24. Vanus J, Bilik P (2022) Research on micro-mobility with a focus on electric scooters within smart cities. World Electric Vehicle J 13:176
25. NOTICE. https://notice.go.jp/en/
26. Yaacoub JPA et al (2022) Robotics cyber security XE "security": vulnerabilities, attacks, countermeasures, and recommendations. Int J Inf Secur 21:115–158

On-Premise Artificial Intelligence as a Service for Small and Medium Size Setups

Carolina Fortuna, Din Mušić, Gregor Cerar, Andrej Čampa, Panagiotis Kapsalis, and Mihael Mohorčič

Abstract Artificial Intelligence (AI) technologies are moving from customized deployments in specific application domains towards generic solutions horizontally permeating vertical domains and industries. For instance, decisions on when to perform maintenance of roads or bridges, or how to optimize public lighting in view of costs and safety in smart cities are increasingly informed by AI models. With new services introduced to smart cities, and the increasing amount of data collected thereof, the in-time decision support will require the transition from developing mostly static AI models to so called AI as a Service (AIaaS) approach. Various commercial solutions, typically deployed in public cloud, already offer user friendly and easy to use AIaaS, functionality-wise enabling the democratization of such ecosystems. However, open-source equivalent ecosystems supporting private deployment for better control over data, processes and costs, where this is needed, are lagging behind, mostly because they require different sets of advanced skills to select and integrate the most suitable technologies for a given application area. In this chapter, we discuss

C. Fortuna · D. Mušić · G. Cerar · M. Mohorčič (✉)
Jozef Stefan Institute, Jamova 39, Ljubljana 1000, Slovenia
e-mail: miha.mohorcic@ijs.si

C. Fortuna
e-mail: carolina.fortuna@ijs.si

D. Mušić
e-mail: din.music@ijs.si

G. Cerar
e-mail: gregor.cerar@ijs.si

D. Mušić
Jozef Stefan International Postgraduate School, Jamova 39, Ljubljana 1000, Slovenia

A. Čampa
Comsensus, Verovškova ulica 64, Ljubljana 1000, Slovenia
e-mail: andrej.campa@comsensus.eu

P. Kapsalis
National Technical University of Athens, Iroon Polytechneiou 9, Attiki 15773, Greece
e-mail: pkapsalis@epu.ntua.gr

R. Shinkuma et al. (eds.), *Advances in Engineering and Information Science Toward Smart City and Beyond*, Engineering Cyber-Physical Systems and Critical Infrastructures 5, https://doi.org/10.1007/978-3-031-29301-6_3

AIaaS functionality and corresponding technology stack and analyze possible realizations using open source user friendly technologies that are suitable for on-premise small and medium sized setups allowing full control over the data and technological platform without any third-party dependence or vendor lock-in.

Keywords AIaaS · Infrastructure automation · Machine learning operations · Technology stack · On-premise

Acronyms

AI	Artificial Intelligence.
AIaaS	Artificial Intelligence as a Service.
API	Application Programming Interface.
CLI	Command Line Interface.
DAG	Directed Acyclic Graph.
DVC	Data Version Control.
HA	High-Availability.
IaaC	Infrastructure as a Code.
KDP	Knowledge Discovery Process.
LKP	libvirt-k8s-provisioner.
ML	Machine Learning.
MLaaS	ML as a Service.
MLOps	ML Operations.
STaaS	Storage as a Service.
XaaS	Anything as a Service.

1 Introduction

Artificial Intelligence (AI) technologies have been mostly used by experts to create new services and manage the extraction of value from large amount of data enabling the rise of some technology giants. However, recently they are moving from customized deployments in specific domains towards more generic solutions aiming for horizontally permeating various vertical domains and industries [13]. For instance, the decision in view of costs and safety in smart cities are increasingly informed by AI models as shown by Martinez et al. [15] on when to perform maintenance of roads or bridges and by De Paz et al. [6] on how to optimize public lighting. However, as argued by Banifatemi et al. [2], to be successful in increasing our overall productivity, the adoption by non-experts in their day-to-day use needs to be further supported by efforts to lower the entry barrier by democratizing AI through the development of intuitive, easy to set up, manage and use systems such as AI as a Service (AIaaS).

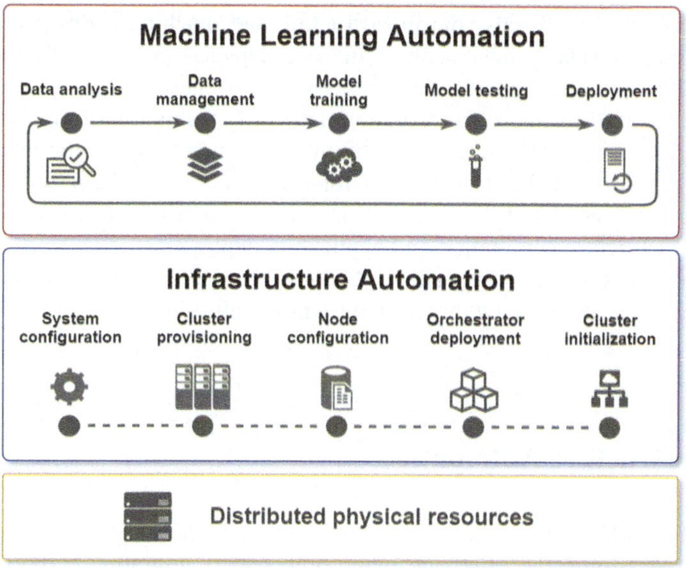

Fig. 1 AIaaS technology stack

Figure 1 presents a three layer abstraction of the AIaaS technology stack consisting of physical resources, infrastructure automation and machine learning automation, while an alternative view of a three layer abstraction focused less on infrastructure automation and more on AIaaS for different target groups is presented in Huang and Rust [11]. The existing cloud service providers initially developed computing and storage functionalities such as Storage-, Infrastructure-, Platform- as-a-Service that were able to abstract physical resources and thus enabled an ecosystem for accelerated application development and scale-up of various companies [9]. Subsequently, they added additional layers of abstraction and functionality leading to the so-called Anything as a Service (XaaS) that enable ever more advanced application development and value creation [10].

Various commercial solutions already offer user friendly and easy to use AIaaS solutions, functionality-wise enabling the democratization of such systems, but their business and technological models rely on controlling parts of the AIaaS technology stack. As per Hon et al. [10], cloud providers mostly still own the physical resources and keep them off the users' premises. However, Wang et al. [18] show that recent commercial offerings are also enabling on-premise infrastructure deployment of so-called edge cloud computing where the physical resources may be on-premises while the infrastructure automation is managed by the provider. Fully controlled on-premises AIaaS stack is still believed to be challenging and costly to deploy, and typically only available to large companies that have sufficient resources to realize it.

In this chapter we discuss the infrastructure and machine learning automation for cloud-native AIaaS stacks such as the one depicted in Fig. 1. We discuss the functionality and corresponding technology stack for these two layers and analyze possible realizations using open source user-friendly technologies that are suitable for on-premise set-ups of small and medium sized environments, allowing users to retain full control over the data and technological platform without any third-party dependence or vendor lock-in. Such set-ups utilize the cloud-native characteristics of both infrastructure automation and application deployment.

This chapter is organized as follows. Section 2 focuses on infrastructure automation, Sect. 3 focuses on machine learning automation while Sect. 4 concludes the chapter.

2 Infrastructure Automation

In the last decade, cloud computing was already used in many businesses, as discussed by Marston et al. [14], and many others were migrating to public clouds (i.e., AWS,[1] GCP[2] and MSA[3]) to reduce operational and management costs [3]. The advantage of such solutions is the simplicity involved in managing the underlying infrastructure where few clicks or few line changes in configuration files reflect in production as depicted on the left in Fig. 2. Alternatively, when public cloud solutions are discouraged for strategic, political or financial reasons, operational costs can also be reduced by adopting open source tools such as OpenStack, Apache Mesos and Eucalyptus or using these to replace proprietary infrastructure management tools where already in place.

Corradi et al. [5] pointed out that the aforementioned open source solutions are well suited for large-scale environments such as data centres, but add unnecessary complexity and overhead for most smaller clusters. For smaller, non-enterprise setups, such as hobby projects, home labs, or small and micro enterprises, it is hard to find a maintained solution that covers day to day needs. For such needs, one approach is to manually set up a cluster as depicted on the right side in Fig. 2. Manual installation requires many steps to successfully set up a cluster. The first three steps that need to be performed only once are related to preparing the hosts by installing and configuring the hypervisors and possibly some other required software as marked with the grey background in Fig. 2. Next follows a series of steps, marked with red under the *Manual* process on the right in Fig. 2, where each virtual machine must be manually created and configured. This process is prone to human errors and can take days or even weeks, depending on the size of the cluster. In case the cluster needs to be upgraded, each node must be manually reconfigured, which takes almost as much time as setting up a new cluster. These steps can be automated, resulting in a

[1] Amazon Web Services, https://aws.amazon.com/.

[2] Google Cloud Platform, https://cloud.google.com/.

[3] Microsoft Azure, https://azure.microsoft.com/.

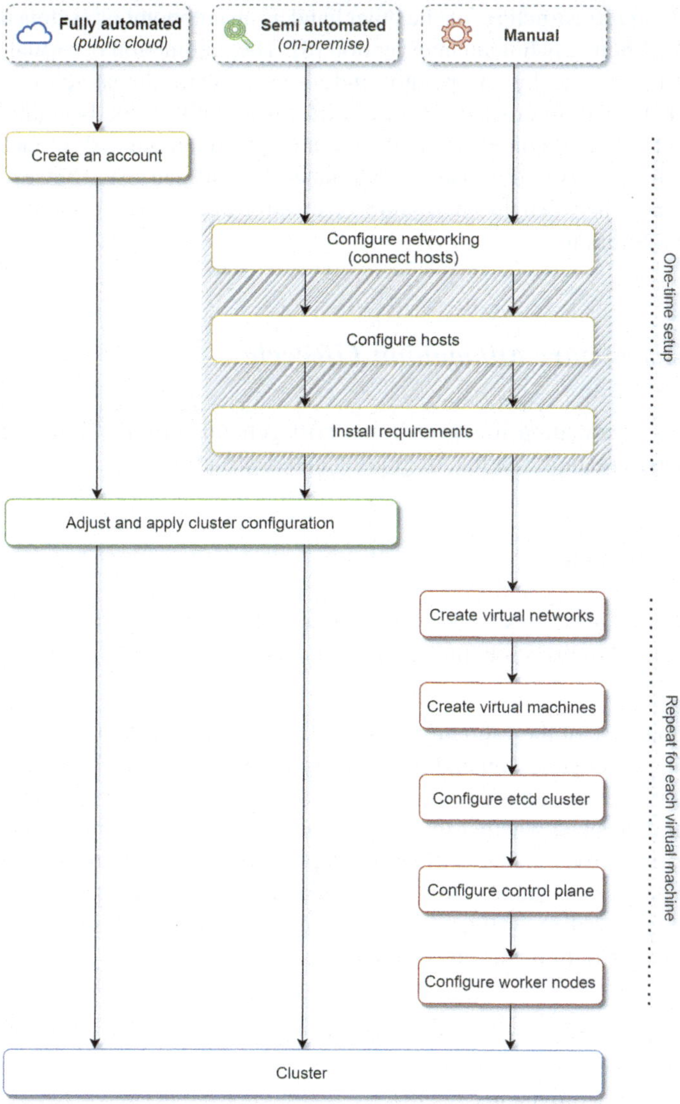

Fig. 2 Comparison of infrastructure deployment steps on fully-automated public cloud, semi-automated on-premise environments and manual installation as perceived by the end user

so-called *Semi-automated* process as illustrated in the middle in Fig. 2, suitable for small to medium size private on-premise setups.

Public cloud infrastructure, on the other hand, is an autonomous system that scales on-demand without any human intervention [8]. The automation of virtualized layer is prepared in advance by an operator and all users share the same infrastructure. As per [4, 12], semi-automated solutions differ from public clouds in that the hosts still need to be preconfigured, while the other steps are automated. In fact, the host configuration process could also be automated by the end user, however, due to large differences between the environments, existing and emerging solutions assume manual configuration.

2.1 Infrastructure Automation Lifecycle

Infrastructure automation lifecycle in its most general form has four independent phases: cluster creation, scaling, upgrading and destruction.

Cluster Creation Phase

As depicted in Fig. 3, cluster creation can be realized through the following four steps: validation, preparation, provisioning and deployment.

Validation
The main objective of the validation step in cluster creation is to detect any potential errors as soon as possible. Depending on the user input, which can be either a terminal command or a set of configuration files, the actions triggered by the input must be predicted and checked whether they can cause potential inconveniences later in the cluster creation process. The validation step is critical to prevent unnecessary or potentially dangerous changes to the system and to save the user's time. Validation of input is followed by validation of system requirements. Based on user input, specific system requirements are verified.

Preparation
The preparation step ensures that missing packages, services, and dependencies are installed. However, conditions that cannot be met during the preparation step must abort the cluster deployment.

Infrastructure Provisioning
After the successful validation and preparation step, the infrastructure provisioning step begins. In this step, virtual components such as the virtual network, virtual machines and storage pools are provisioned according to the user's input. On each virtual machine an operating system is deployed and configured to make it accessible on the local network.

Platform Deployment
The provisioned infrastructure consists of many independent virtual components that are difficult to manage and control. Therefore, a platform has to be deployed

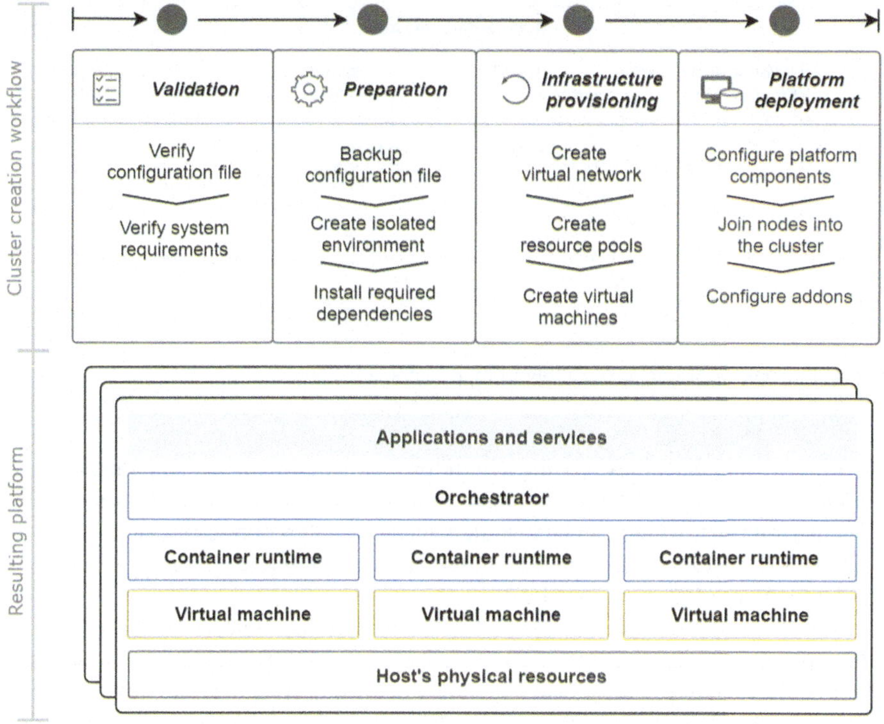

Fig. 3 Infrastructure automation workflow and scheme of the resulting infrastructure

on top of the provisioned virtual components to abstract them into a cluster that is managed as a single entity. Following the cloud-native programming paradigm, the platform must enable autonomous management of microservice-oriented workloads. A microservice can be described as an independent software component that solves a very specific problem. Therefore, multiple loosely coupled microservices communicate with each other and perform as a single cloud-native application. Such an approach allows platforms to automatically scale specific parts of an application as demand increases or decreases, and recover them in the event of failure.

Platform deployment step is responsible for the installation of orchestrator and platform components, and connection of all nodes into the cluster. In this step, all orchestrator components, such as scheduler, server Application Programming Interface (API), and other system controllers, are deployed on each provisioned virtual machine. In addition to traditional virtualization, containerization can also be used. Containerization is a popular, lightweight form of virtualization. It takes advantage of the host's operating system kernel to create isolated processes in different system namespaces. Compared to virtual machines, containers have low overhead because they do not require a dedicated operating system, making container creation, migration and deletion very efficient. By combining containerization with traditional virtu-

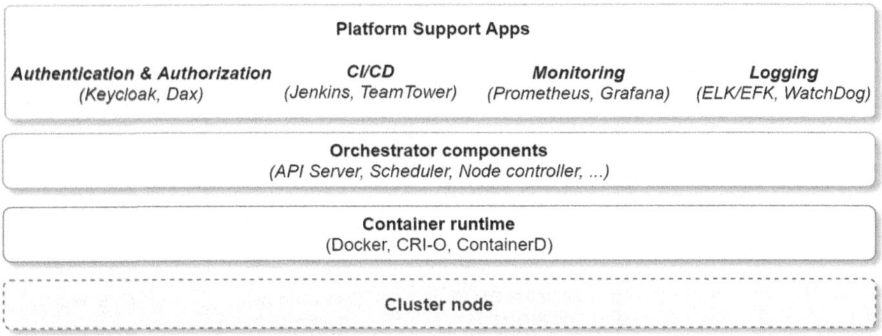

Fig. 4 Example infrastructure resulting from the automation process

alization, virtual machines provide a high level of isolation, while containers reduce workload overhead and facilitate cluster management. Containerization requires a container runtime, which can be installed and configured during this step.

After successful cluster installation, the cluster, as depicted in Fig. 4, is ready for the deployment of custom workload.

In addition to the four general steps, infrastructure automation may also include other optional steps. For instance, the installation of certain applications and services required on each deployed cluster, such as monitoring, logging, and authentication services, marked as *Platform Support Applications* in Fig. 4, can be performed as part of infrastructure automation. However, the deployment of custom workloads, such as the ML Operations (MLOps) pipeline, is typically separated from infrastructure automation to ensure loose coupling of the architecture.

Cluster Scaling, Upgrading and Destruction Phases

In addition to the initial setup of the cluster, the infrastructure automation strategy must also include the subsequent management of the cluster, such as scaling and upgrading the cluster. As the workload increases or decreases, a running cluster may reach its capability limits or use its resources inefficiently. Therefore, scaling the cluster is an important part of the automation process. Cluster scaling consists of adding and removing cluster nodes. Scaling can be triggered either automatically based on collected cluster metrics or manually. In larger environments consisting of many clusters, automatic scaling enables better overall utilization of resources. Typically, resource usage of different clusters varies widely, and the ability to free up unused resources and make them available for other clusters to consume, reduces the required physical resources. However, auto-scaling requires additional components to manage the infrastructure, which increases the complexity of the environment. Therefore, for smaller environments that have more predictable resource consumption, manually triggered scaling may be sufficient.

The process of adding nodes to the cluster is similar to the initialization of the cluster. The additional virtual machines are created and configured first and then the platform components are installed on them. When the cluster nodes are ready, they are simply joined to the existing cluster. Removing nodes, on the other hand, requires more attention because these nodes are already in operation and may contain some active workloads. To gracefully remove a node, the node must first be drained, which means moving all of the workload to other running nodes. This ensures that services remain available after the node is removed.

In addition to cluster scaling, the cluster must be regularly maintained to ensure that all components are up to date. Compared to cluster scaling, upgrades still need to be automated, but are usually triggered manually. This is because the compatibility of newer components must be checked in advance, otherwise the upgrade could render the entire cluster unusable. There are many strategies for upgrading the cluster, but in-place upgrades are most common. With in-place strategy, every node in the cluster is upgraded sequentially one-by-one. This is done by draining the node, upgrading its components and rejoining the node back to cluster. No additional virtual components need to be created for this approach, which is why it is frequently used, especially in smaller environments.

Once the cluster reaches the end of its life, it can be removed. To remove the cluster completely, only the virtual components need to be stopped and deleted. However, if the cluster contains valuable data, the data must be safely migrated before the cluster is destroyed, because it will be lost during this process.

2.2 Technology Stack for On-Premise Infrastructure Automation

Figure 5 shows the most popular open source technologies commonly used for infrastructure automation. A sensible combination of such technologies can be employed to realize an on-premises infrastructure according to the example depicted in Fig. 4.

With respect to provided functionalities, technologies are divided into 4 groups comprising of Virtualization, Containerization, Orchestration and Automation Tools. Virtualization technologies represent hypervisors that are essential for traditional virtualization on top of physical resources. They are capable of running multiple guest operating systems on a single physical host while isolating them from each other. KVM[4] and Xen[5] are open source projects, with KVM being the most widely used. Hyper-V[6] and VMWare[7] are good examples of proprietary alternatives.

---

[4] KVM, https://www.linux-kvm.org.

[5] Xen, https://xenproject.org.

[6] Hyper-V, https://learn.microsoft.com/en-us/windows-server/virtualization/hyper-v/hyper-v-technology-overview.

[7] VMWare solutions, https://www.vmware.com.

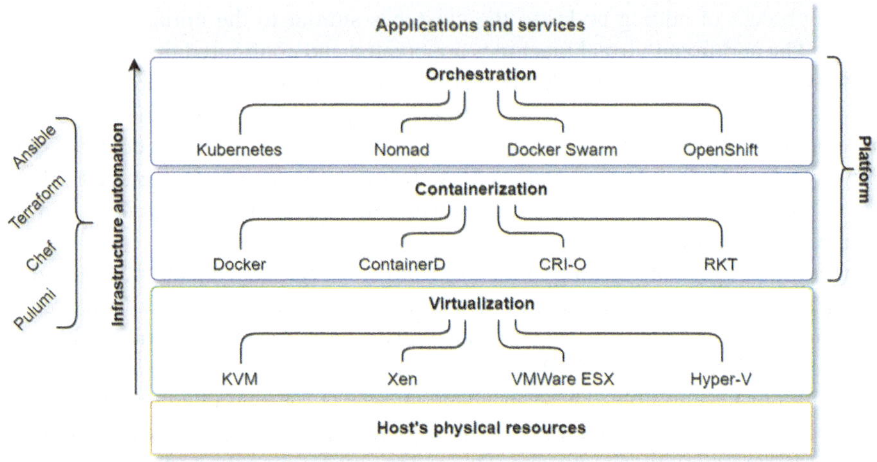

Fig. 5 Infrastructure technology stack and popular automation tools

The second group contains technologies for containerization. Just as traditional virtualization requires a hypervisor to manage virtual resources, container runtimes enable containerization. The most popular container runtimes are Docker,[8] ContainerD,[9] CRI-O,[10] and RKT.[11]

Container orchestrators can efficiently coordinate applications packaged as containers across multiple containerized environments. The most popular container orchestrator is Kubernetes,[12] which also supports most of the container runtimes mentioned above. Nomad[13] and OpenShift[14] are seen as good alternatives for Kubernetes. Docker Swarm,[15] on the other hand, offers native support for orchestrating Docker environments. However, it cannot be used with any other container runtime.

Finally, automation tools such as Ansible,[16] Terraform,[17] Pulumi,[18] and Chef[19] are used for provisioning a set of virtual machines, deploying container runtimes, and installing various dependencies. All these Infrastructure as a Code (IaaC) tools belong to the declarative languages, in which the desired state of the infrastruc-

[8] Docker, https://www.docker.com/.

[9] ContainerD, https://containerd.io.

[10] CRI-O, https://cri-o.io.

[11] RKT, https://www.redhat.com/en/topics/containers/what-is-rkt.

[12] Kubernetes, https://kubernetes.io.

[13] Nomand, https://www.nomadproject.io.

[14] OpenShift, https://www.redhat.com/en/technologies/cloud-computing/openshift.

[15] Docker Swarm, https://docs.docker.com/engine/swarm.

[16] Ansible, https://www.ansible.com.

[17] Terraform, https://www.terraform.io.

[18] Pulumi, https://www.pulumi.com.

[19] Chef, https://www.chef.io.

ture is described rather than precisely defined with specific steps. However, typical imperative languages such as Go and Python can be used to achieve the same result. Imperative languages offer more freedom, but the automation process may take longer.

2.3 Semi-automated Solutions for On-Premise Infrastructure Deployments

In Table 1 we have identified open source solutions that can be used to set up on-premise clusters. These solutions make use of the technologies discussed above and depicted in Fig. 5.

All solutions analyzed in Table 1, Kubitect,[20] libvirt-k8s-provisioner (LKP),[21] KubeNow,[22] MicroK8s,[23] and K3s,[24] are classified as semi-automated because they have some unique requirements that must be met by the user. Kubitect and LKP are capable of creating virtualized resources as part of the automation process. MicroK8s and K3s can also be used on top of the traditional virtualization, however, they do not provide infrastructure automation as part of the solution. Thus, without additional automation, MicroK8s and K3s nodes must be manually configured on each host. KubeNow differs from other solutions as it can only be used with OpenStack, an open source private cloud solution, for on-premises deployments.

Kubitect, KubeNow and MicroK8s clusters are created and managed via custom Command Line Interface (CLI) tools of the corresponding solutions. K3s and LKP clusters, on the other hand, are managed directly via scripts, which are less intuitive to use.

All solutions use Kubernetes as an application and service orchestrator and, with the exception of KubeNow, primarily use ContainerD as a container runtime environment. While KubeNow only supports Docker, LKP can be configured to use either Docker, CRI-O or ContainerD runtimes. However, Docker and ContainerD are de facto standards and should cover most users' needs.

In terms of implementation, each solution uses a different set of technologies to create Kubernetes clusters. The Kubitect CLI tool is written in Go and further takes advantage of Terraform and Ansible for infrastructure deployment and configuration. LKP uses only Terraform and Ansible. Consequently, different scripts must be run to create or destroy the cluster. MicroK8s and K3s, on the other hand, do not use Ansible or Terraform, but they also do not provide automation of the virtualized infrastructure. Instead, the entire solution is implemented using a single programming language. MikroK8s is written in Python and K3s in Go.

---

[20] Kubitect, https://kubitect.io.

[21] libvirt-k8s-provisioner, https://github.com/kubealex/libvirt-k8s-provisioner.

[22] KubeNow, https://github.com/kubenow/KubeNow.

[23] MicroK8s, https://microk8s.io.

[24] K3s, https://k3s.io.

Table 1 Comparison of semi-automated infrastructure automation solutions for small and medium set-ups

Projects\ Parameters	Kubitect	LKP	KubeNow	MicroK8s	K3s
On-premise traditional virtualization	Libvirt	Libvirt	OpenStack	/	/
Management tool	CLI tool (kubitect)	Ansible scripts	CLI tool (kn)	CLI tool (microk8s)	Shell scripts
Supported container runtimes	ContainerD	ContainerD, CRI-O, Docker	Docker	ContainerD	ContainerD
Container orchestrator	Kubernetes	Kubernetes	Kubernetes	Kubernetes	Kubernetes
Multi-host deployments	Y	N	Y	Y	Y
Cluster scaling	Y	N	Y	Y	Y
Cluster upgrades	Y	N	N	Y	Y
High availability (HA) cluster topology	Y	N	N	Y	Y
Out of the box storage solution	Rook	Rook	*Provider dependent*	/	Longhorn

Besides the implementation details, the solutions differ mainly in the supported topologies and management features. LKP does not support cluster deployments across multiple physical hosts, which means that it is not possible to create truly high-availability (HA) clusters. Likewise, it does not provide any other management functions, such as scaling and upgrading existing clusters. Kubitect and KubeNow, on the other hand, support cluster deployment across multiple hosts. However, only Kubitect is capable of creating HA clusters, as KubeNow is limited to a single control plane node whose failure renders the entire cluster unusable. MicroK8s and K3s also support cluster deployments that span multiple hosts. However, since they do not have infrastructure automation out of the box, this must be done manually.

Every created cluster also requires a place to store the application's data. While most storage solutions are deployed on top of the Kubernetes orchestrator and are able to consume disks attached to the host, some solutions provide configuration of storage clusters out of the box. Kubitect and LKP both provide optional setup of distributed storage cluster that uses Rook as the storage orchestrator. Using Rook,[25] different kinds of storage types can be used, such as filesystem, object store and block storage, which covers all requirements of MLOps. MicroK8s does not provide a distributed storage solution out of the box, neither as part of the cluster deployment nor as an optional add-on. In comparison, K3s offers the Longhorn[26] deployment as an optional add-on module that is equivalent to Rook in terms of features. KubeNow is only able to use the storage provided by the underlying cloud.

[25] Rook, https://rook.io.

[26] Longhorn, https://longhorn.io.

3 Machine Learning Automation

Assuming the on-premise infrastructure is provided, as described in Sect. 2, we can proceed with the deployment of the AIaaS. The development of machine-learning models follows a well defined Knowledge Discovery Process (KDP) [7], which consists of the following four steps: (1) data analysis, (2) data preparation (pre-processing), (3) model training and evaluation, and (4) model deployment [17]. In the past, such process involved intense manual effort and the enabling tools were familiar only to a limited number of domain experts. However, in recent years, coordinated efforts have been taken by the private and public sectors to democratize AI and model development [1] to empower also less specialized users.

The democratization process involves a division of labour and automation approach applied to the KDP, as elaborated more in details by Ruf et al. in [17], where rather than a domain expert executing the process in Fig. 6 step-by-step, the process only needs to be controlled at a few key steps. Furthermore, Huang and Rust in [11] suggested that generic models can also be trained and made available through inference API where the users only leverage an existing, pre-trained model to make an inference used by their application or service. This process can be further automated by MLOps [17].

MLOps automate the process of building, testing, and deploying machine learning models. As depicted in Fig. 6, MLOps process can be represented with the following four phases: (1) data ingestion and purification, (2) feature management, (3) model development, and (4) model management.

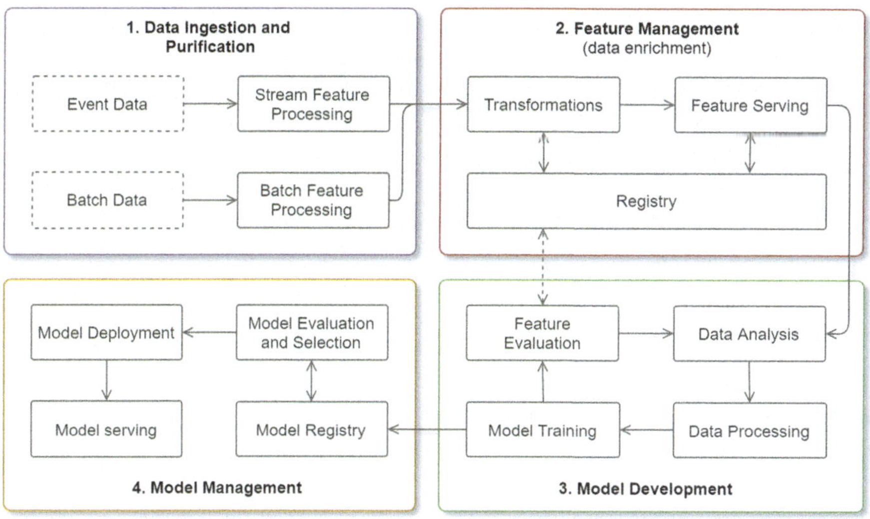

Fig. 6 Machine learning automation workflow

In the Data Ingestion and Purification phase, data from sensors or applications can be received in batch format such as hourly, daily or weekly batches or in a streaming format as it is being produced. The incoming data is then managed by feature stores rather than manually through generation scripts and file/database storage. Feature stores are data management services that harmonize data processing steps in producing features for different machine learning (ML) pipelines, making it more cost-effective and scalable compared to traditional approaches [16], and they can also be seen and made available as a service (i.e., Storage as a Service (STaaS)).

During the model development phase, various combinations of features available in the feature store are used to train ML models. The model training step can be made available to ML developers as ML as a Service (MLaaS), where they can tune and customize desired models on desired features [11], or they can be entirely abstracted and hidden from less advanced users. The resulting models are stored in a registry, evaluated and some of them eventually deployed as part of the Model Serving step. The model evaluation and deployment enable the development of the inference API services where less experienced users are able to select and use pre-trained models for their applications.

3.1 *Technology Stack for Cloud-Native ML Automation*

The choice of technologies available for realizing the ML automation workflow depicted in Fig. 6 is becoming richer every year. In Fig. 7, we provide an overview of possible candidates. During the data ingestion phase, a batch can, for instance, be

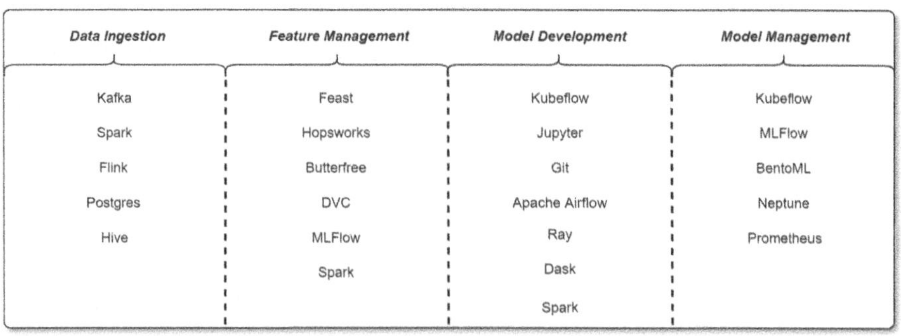

Fig. 7 MLOps stack for on-premises deployment

Table 2 List of open-source feature store solutions suitable for time series data

Name	Open source	Data sources	Offline storage	Online storage	Deployment
Feast	Y	BigQuery, Hive, Kafka, Parquet, Postgres, Redshift, Snowflake, Spark, Synapse	BigQuery, Hive, Pandas, Postgres, Redshift, Snowflake, Spark, Synapse, Trino, custom	DynamoDB, Datastore, Redis, Azure Cache for Redis, Postgres, SQLite, custom	AWS Lambda, Kubernetes, local
Hopsworks	Y	Flink, Spark, custom Python, Java, or Scala connectors	Azure Data Lake Storage, HopsFS, any SQL with JDBC, Redshift, S3, Snowflake	any SQL with JDBC, Snowflake	AWS, Azure, Google Cloud, local
Butterfree	Y	Kafka, S3, Spark	S3, Spark Metastore	Cassandra	Local

retrieved from a database or STaaS service (e.g., BigQuery,[27] Hive,[28] PostgreSQL,[29]) or a stream processing platform (e.g., Kafka,[30] Spark[31] Flink[32]).

For managing the features, fully automated solutions such as feature stores (e.g., Feast,[33] Hopsworks[34]) can be employed or it can also be realized using data versioning and data processing solutions such as Data Version Control (DVC),[35] Pandas,[36] and Spark. Several closed source feature stores have been recently developed, while to the best of our knowledge, only three feature management solutions are available as open source with their characteristics summarised in Table 2. It can be seen in the third column of the table that they already include connectors to support fast interconnection with various storage solutions (e.g., BigQuery, S3,[37] PostgreSQL) and streaming platforms (e.g., Kafka, Spark). As per columns four and five, it can be seen that all open source feature stores support offline and online storage such as public cloud provider's BigQuery, Azure, S3 and Snowflake[38] or open source solu-

[27] Google BigQuery, https://cloud.google.com/bigquery.

[28] Apache Hive, https://hive.apache.org.

[29] PostgreSQL, https://www.postgresql.org.

[30] Apache Kafka, https://kafka.apache.org.

[31] Apache Spark, https://spark.apache.org.

[32] Apache Flink, https://flink.apache.org.

[33] Feast, https://feast.dev.

[34] Hopsworks, https://www.hopsworks.ai.

[35] DVC, https://dvc.org.

[36] Pandas, https://pandas.pydata.org.

[37] Amazon S3, https://aws.amazon.com/s3.

[38] Snowflake, https://www.snowflake.com.

tions such as PostgreSQL and Cassandra.[39] As can be seen from the sixth column of the table, the open source feature stores can be deployed locally in an on-premise setup and also in the public cloud.

For the model development phase, the necessary tools include tools that enable code development and versioning such as Kubeflow,[40] Jupyter[41] and Git[42]; tools that are responsible to orchestrate model training such as Kubeflow Pipelines, Airflow[43] and CML[44]; as well as tools for speeding up training through parallelization such as Ray,[45] Dask[46] and Spark. Using the set of tools in this phase, MLaaS can be realized where developers are presented with a complete playground for custom model development.

Finally, the model management phase comprises of tools for model tracking, evaluation and serving such as Kubeflow and MLFlow.[47] Especially these two tools are evolving rapidly with largely overlapping functionality in the model development and model management phases of MLOps. BentoML,[48] KServe,[49] Seldon Core,[50] MLEM,[51] TensorFlow Serving[52] and Kale[53] can also be used for model serving, *Neptune.ai*[54] or *"Weights & Biases"*[55] for versioning while Prometheus[56] for model performance monitoring.

3.2 On-premise AIaaS Deployment Guidelines

In Sect. 3.1, we analyzed the diverse range of technologies and solutions that are available for various phases of MLOps, as depicted in Fig. 6. While each solution addresses a subset of issues within the MLOps ecosystem and offers high levels of flexibility, this can be overwhelming for new adopters. Therefore, in this section,

---

[39] Apache Cassandra, https://cassandra.apache.org.

[40] KubeFlow, https://www.kubeflow.org.

[41] Jupyter, https://jupyter.org.

[42] GIT, https://git-scm.com.

[43] Apache Airflow, https://airflow.apache.org.

[44] CML, https://cml.dev.

[45] Ray, https://www.ray.io.

[46] Dask, https://www.dask.org.

[47] MLFlow, https://mlflow.org.

[48] BentoML, https://www.bentoml.com.

[49] KServe, https://kserve.github.io.

[50] Seldon Core, https://www.seldon.io/solutions/open-source-projects/core.

[51] MLEM, https://mlem.ai.

[52] TensorFlow Serving, https://www.tensorflow.org/tfx.

[53] KubeFlow Kale, https://www.kubeflow.org/docs/external-add-ons/kale.

[54] Neptune.ai, https://neptune.ai.

[55] Weights & Biases, https://wandb.ai.

[56] Prometheus, https://prometheus.io.

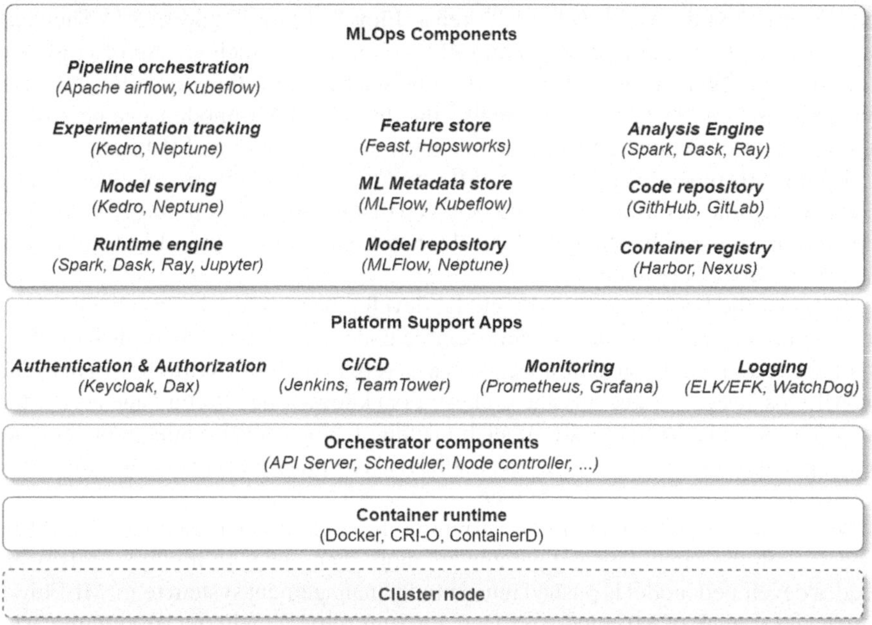

Fig. 8 Example on-premise infrastructure for AIaaS

we outline the phases and steps depicted in Fig. 6 and, based on our experiences, present a selection of open-source solutions that work together well and is almost ML framework agnostic.

Figure 8 shows an example of an ML automation platform deployed on top of on-premise semi-automated infrastructure. To build a model, we first require data. A feature store (e.g., Feast) can be used to deal with live/streaming sensor data (i.e., events) from a stream processor (e.g., Kafka) and historical data from different data sources (e.g., data warehouse/lake). A registry within a feature store contains instructions for obtaining or composing features. With all the instructions set and all data sources available, the feature store unifies access to the features both for production and development. Once a new instruction is added to the feature store's registry, a new feature becomes immediately available.

The model development phase depicted in Fig. 6 focuses on analyzing data and building ML models with ML pipelines. Data analysis and model experimentation can lead to a discovery of new features that can be valuable addition for future models. Therefore, they are contributed to the feature store (see above) as instructions in the registry. Machine learning pipelines, which define process from data to machine learning model, can be specified in code (e.g., Python script, Jupyter Notebook) using open-source machine learning frameworks such as NumPy,[57] SciPy,[58]

[57] NumPy, https://numpy.org.

[58] SciPy, https://scipy.org.

scikit-learn,[59] XGBoost,[60] PyTorch,[61] TensorFlow,[62] Keras,[63] and JAX.[64] They can be extended with higher-order "AutoML" frameworks, such as AutoKeras[65] or AutoSklearn,[66] which automate search for optimal ML algorithms, architectures, and hyper-parameters for given data. Finally, the developed ML models can be pushed automatically or manually to a model management system (e.g., MLFlow).

Tuning ML pipeline and therefore ML models is a complex process that usually requires multiple iterations. It should be noted that the feature importance in MLOps pipelines is assessed by the model development phase depicted in Fig. 6 subject to the availability of the features.

To make the best out of an ML model development process, we consider good practice to keep track of ML pipeline source code with versioning tools (e.g., Git) and have it co-located on centralized services (e.g., GitHub,[67] GitLab[68]). This way, the MLOps system (especially the orchestrator) knows where to find and access the latest source code. With a source code branching feature of versioning tools, we can develop features or improvements in parallel and test them independently before they get merged into the production branch.

ML pipeline steps can be time-consuming. It would be inefficient to rebuild ML models with the same data and parameters more than once. To minimize this overhead, a developed model is pushed into a model management system (e.g., MLFlow), which keeps track of ML model artifacts for later retrieval without (re)training. The artifacts are, for instance, model hyper-parameters, self-evaluation metrics, model's score, and model's binary representation (i.e., inner state, model parameters, model weights). The model registry within a model management system keeps the pipeline source code and artifacts connected by preserving essential parts of the source code and source code commit hash.

Once models are preserved in a model management system (i.e., MLFlow), they can be accessed and inspected through the web or API. Individual models with their meta-data, input parameters, and evaluation metrics can be inspected and compared with other models. These models can be manually (via web interface) or automatically (through API) promoted or demoted into *test* or *production* stage, once they undergo additional inspection and evaluation.

During the model deployment step, the automated pipeline takes models in *test* or *production* stage, converts them into containers suitable for the model serving

---

[59] scikit-learn, https://scikit-learn.org.

[60] XGBoost, https://xgboost.readthedocs.io/en/stable.

[61] PyTorch, https://pytorch.org.

[62] TensorFLow, https://www.tensorflow.org.

[63] Keras, https://keras.io.

[64] JAX, https://github.com/google/jax.

[65] AutoKeras, https://autokeras.com.

[66] AutoSklearn, https://automl.github.io/auto-sklearn.

[67] GitHub, https://github.com.

[68] GitLab, https://gitlab.com.

framework (e.g., MLFlow) and prepares them for deployment. Once models are deployed by MLFlow, MLOps pipeline exposes their interaction (i.e., inference) interfaces.

For orchestrating, automating, and synchronizing MLOps pipelines, Apache Airflow can be used. The pipeline orchestrator requires instructions made out of small tasks, which can depend on each other. However, their inter-dependencies must form a directed acyclic graph (DAG). The pipeline orchestrator can also be used for preparing containers, rebuilding models, and (re)deploying models.

In general, an on-premise AIaaS deployment consists of many complex interconnecting services such as the Apache Airflow orchestrator, Feast feature store, JupyterHub, MLFlow model storage and RedisDB object storage. However, most of these services consist of many smaller hidden building blocks. For example, MLFlow model storage requires an MLFlow server, a PostgreSQL database for storing models' metadata, and MinIO for storing models' artifacts. To reduce the complexity, the usual practice is to containerize individual building blocks and split applications into multiple microservices.

4 Conclusions

In this chapter, we discussed the concept of AIaaS along with recent developments in automation that enable on-premises AIaaS deployments, thus extending the benefits of such systems also to small and medium size setups requiring full control over the data and technological platform. We first focused on the general process required for on-premise infrastructure deployment and identified a number of existing open source tools and technologies for possible realizations. Then we discussed the general process required for machine learning pipeline automation in view of enabling AIaaS and proceeded with the identification of suitable technologies for its on-premise implementation.

Overall, we argue that the available open source technologies are sufficiently mature and suitable for small to medium size on-premise AIaaS setups. They can be automatically configured and interconnected in such a way as to support easy and intuitive setting up, use and management, while hiding the complexity from the non-expert users. Such AIaaS setups can thus support gradual introduction of smart services also to various stakeholders and organizational entities in cities and communities of various sizes, and enable their extension and scaling with the increasing needs and introduction of new services and applications.

Most of the open-source solutions for on-premise AIaaS systems described in this chapter are building blocks that address a particular subset of problems within the MLOps ecosystem. While this offers high flexibility, it often leads to reinventing their integration because the existing MLOps services are only partially interchangeable. In addition, it is also challenging to migrate stateful parts of the stack, such as databases, feature registries, and model registries. Choosing the right technology stack and migration represents an important challenge for deploying on-premise

AIaaS systems and requires some expert knowledge, at least for the initial setup. In this respect, a possible future work is to create a turn-key solution that can be easily deployed and configured in an on-premise environment by less experienced users. This involves the development of a pre-configured cluster or machine along with all necessary components for an AIaaS platform, including a machine learning platform, model serving infrastructure as well as data storage and management tools.

Another possible direction for future work is to create a maintained AIaaS solution based on open-source components that can be customized and configured to meet the specific needs of an enterprise. This would involve building and offering a suite of modular, interoperable components that can be easily integrated and configured to support different use cases and workflows. In this approach, the use of open-source components could notably reduce the costs and facilitate faster onboarding of new developers.

In summary, the efforts for simplified deployment of on-premise AIaaS solutions are believed to improve the acceptability of AI in society by reducing the entry barrier for organizations and individuals to utilize the technology and taking advantage of its benefits. In addition, on-premise environments provide greater control over data and its use, thus addressing the concerns around privacy and security of AI and cloud services.

Acknowledgements This work was funded in part by the Slovenian Research Agency under the grant P2-0016 and the European Commission under grant agreements 101000158 (MATRYCS) and 872525 (BD4OPEM).

References

1. Allen B, Agarwal S, Kalpathy-Cramer J et al (2019) Democratizing AI. J Am Coll Radiol 16(7):961–963
2. Banifatemi A, Miailhe N, Buse Çetin R, et al (2021) Democratizing AI for humanity: a common goal. In: Reflections on artificial intelligence for humanity. Springer, pp 228–236
3. Bibi S, Katsaros D, Bozanis P (2012) Business application acquisition: on-premise or saas-based solutions? IEEE Softw 29(3):86–93
4. Capuccini M, Larsson A, Carone M et al (2019) On-demand virtual research environments using microservices. PeerJ Comput Sci 5:e232
5. Corradi A, Fanelli M, Foschini L (2014) Vm consolidation: a real case based on openstack cloud. Future Generat Comput Syst 32:118–127. https://doi.org/10.1016/j.future.2012.05.012, https://www.sciencedirect.com/science/article/pii/S0167739X12001082, Special Section: The Management of Cloud Systems, Special Section: Cyber-Physical Society and Special Section: Special Issue on Exploiting Semantic Technologies with Particularization on Linked Data over Grid and Cloud Architectures
6. De Paz JF, Bajo J, Rodríguez S et al (2016) Intelligent system for lighting control in smart cities. Inf Sci 372:241–255
7. Fayyad UM, Piatetsky-Shapiro G, Smyth P, et al (1996) Knowledge discovery and data mining: towards a unifying framework. In: KDD, pp 82–88
8. Garrison J, Nova K (2017) Cloud native infrastructure: patterns for scalable infrastructure and applications in a dynamic environment. O'Reilly Media, Inc.

9. Goyal S (2014) Public vs private vs hybrid vs community-cloud computing: a critical review. Int J Comput Netw Inf Sec 6(3):20–29
10. Hon WK, Millard C, Singh J (2022) Cloud computing demystified (part 1): technical and commercial fundamentals. Available at SSRN 4030064
11. Huang MH, Rust RT (2018) Artificial intelligence as a service. J Serv Res 21(2):155–172
12. libvirt-k8s-provider (2022) Automate your cluster provisioning from 0 to k8s! https://github.com/kubealex/libvirt-k8s-provisioner, Accessed on 26 Apr 2022
13. Makridakis S (2017) The forthcoming artificial intelligence (AI) revolution: its impact on society and firms. Futures 90:46–60
14. Marston S, Li Z, Bandyopadhyay S et al (2011) Cloud computing - the business perspective. Decis Support Syst 51(1):176–189. https://doi.org/10.1016/j.dss.2010.12.006, www.sciencedirect.com/science/article/pii/S0167923610002393
15. Martinez P, Mohamed E, Mohsen O, et al (2020) Comparative study of data mining models for prediction of bridge future conditions. J Perform Constr Facil 34(1):04019,108
16. Patel J (2020) Unification of machine learning features. In: 2020 IEEE 44th annual computers, software, and applications conference (COMPSAC), pp 1201–1205. https://doi.org/10.1109/COMPSAC48688.2020.00-93
17. Ruf P, Madan M, Reich C et al (2021) Demystifying mlops and presenting a recipe for the selection of open-source tools. Appl Sci 11(19):8861
18. Wang W, Tornatore M, Zhao Y, et al (2021) Infrastructure-efficient virtual-machine placement and workload assignment in cooperative edge-cloud computing over backhaul networks. IEEE Trans Cloud Comput 1–1. https://doi.org/10.1109/TCC.2021.3107596

Incentives in Surplus Food Distribution for Smart Cities and Beyond: An Activity Aware Solution

Surja Sanyal, Sajal Mukhopadhyay, Fatos Xhafa, Vikash Kumar Singh, and Banhi Sanyal

Abstract Food surplus donation has emerged as a popular means to counter the food wastage issue that the world has been facing in recent years. The fact that this also addresses the food insecurity challenge is an added bonus. The matching of the surplus donors with the receivers needs the constant availability of willing volunteers to make this movement a success. An incentive module attracts volunteer participation and ensures their constant availability for the smooth movement of surplus food. While the existing incentive mechanism gets the attention of the volunteers, it fails to account for the degree of activity of the volunteers. This work modifies the incentive scheme to ensure that the activities of the volunteers are taken into consideration. The proposed mechanism manages to discourage inactivity with lower incentives and promote proactivity with higher incentives while disbursing the same total incentive amount taking all volunteers together.

S. Sanyal
Department of Computer Science and Engineering, Indian Institute of Technology Guwahati, Guwahati, Assam, India
e-mail: hi.surja06@gmail.com

S. Mukhopadhyay
Department of Computer Science and Engineering, National Institute of Technology Durgapur, Durgapur, West Bengal, India
e-mail: sajal@cse.nitdgp.ac.in

F. Xhafa (✉)
Department of Computer Science, Universitat Politècnica de Catalunya, Barcelona, Spain
e-mail: fatos@cs.upc.edu

V. Kumar Singh
School of Computer Science and Engineering, VIT-AP University, Amaravati, Andhra Pradesh, India
e-mail: vikash.singh@vitap.ac.in

B. Sanyal
School of Computer Engineering, Kalinga Institute of Industrial Technology, Bhubaneswar, Odisha, India
e-mail: banhisanyal9@gmail.com

R. Shinkuma et al. (eds.), *Advances in Engineering and Information Science Toward Smart City and Beyond*, Engineering Cyber-Physical Systems and Critical Infrastructures 5, https://doi.org/10.1007/978-3-031-29301-6_4

Acronyms

AAIM	Activity Aware Incentive Mechanism.
FDRM-CA	Food Donor to Receiver Matching with Chronological Acceptance.
ICT	Information and Communications Technology.
IoT	Internet of Things.

Glossary

Activity based rewarding	The reward amount is directly proportional to the level of participation of the volunteer in surplus movement.
Chain of volunteers	The contiguous list formed by the volunteers present in the shortest path of the delivery person.
Delivery person	The volunteer who actually delivers a surplus food donation.
Donor-receiver relations	Informal relations that build up due to interactions for surplus movement.
Expected payment	The expected earnings of any arbitrary volunteer.
Exploit and explore	The concept of using the higher rewarding option more frequently, and using the unknown options less frequently only to find a more rewarding option.
Exploitable volunteers	Those volunteers who are actively working and will be rewarded by the system more frequently so as to motivate them to participate more.
Exploitation	The more frequent utilization of a highly rewarding option.
Exploration	The less frequent utilization of unknown options in search of an option that rewards higher than the current highest rewarding one.
Flat rewarding	The reward amount does not consider the level of participation of the volunteer in surplus movement.
Food habits	The practices of an individual that relate to the production, handling, and consumption of food.
Food supply chain	The processes responsible for bringing food from the fields to our plate.
Food insecurity	Inability to get regular access to enough safe and nutritious food for normal growth and development.
Food wastage	Food that is consumable but is consciously discarded.

Hyperparameter	The metadata variable of a mechanism which has a direct effect on the behaviour of the mechanism.
Identity duplication	Using multiple identities to obtain multiple resources/services.
Inactivity penalty	The less incentive that a volunteer gets when the volunteer does not participate actively in surplus movement.
Incentive mechanism	The mechanism that runs under the hood of the incentive module.
Incentive module	The module of the system that calculates and disburses incentives to the volunteers.
Invitations	Volunteers inviting other probable volunteers to the system.
Matching module	The module of the system that matches donors/donations with receivers/receipt requests through volunteers.
Mobility of the surplus	The pickup and delivery of the surplus food from its donor to its receiver.
Multi-armed bandit mechanism	The mechanism that maximizes rewards by utilizing the currently highest rewarding option more frequently, and exploring the unknown options with less frequency in search of a more rewarding option.
Operational practices	The methods that are used to operate a service/business.
Operational identity	The identity of a volunteer used to register and execute the surplus food donation requests.
Parent-children construct	The one to many hierarchy formed by volunteers consisting of one inviter (parent) and many invitee (children) volunteers.
Passive income	Income received by a volunteer generated by the work of other volunteers.
Perishable food items	Food items that spoil quickly and need immediate refrigeration.
Proactivity bonus	The extra incentive that a volunteer gets when the volunteer participates actively in surplus movement.
Probabilistic selection	The selection of volunteers based on probability/chance.
Scalable	Capable of change, often increase, in size.
Shift of funds	The increase or decrease of individual volunteer incentive amounts.
Shortest path	The route from the delivery person to the volunteer directly attached to the system without visiting any volunteer twice.

Smart cities	A municipality that uses Information and Communications Technology (ICT) for increased efficiency, information exchange, and public and government welfare.
Stabilize	Dampening of the fluctuations in the availability of volunteers.
Surplus food distribution	The redistribution of the extra food that was prepared, but is not being consumed.
Surplus food	The excess food that was prepared, but is not being consumed.
Sustainability	The ability to stay afloat in the long run.
Sybil attacks	To obtain multiple resources/services by operating from multiple identities so as to starve others from their rightful resources/services.
Technological platforms	The Information and Communications Technology (ICT) and Internet of Things (IoT) based platforms.
Top performers	Those volunteers whose participation in surplus distribution is more than some predefined threshold level.
Total incentive disbursed	The cumulative incentive disbursed by the system for one successful surplus delivery.
Volunteer participation	The measure of active surplus distribution by volunteers.
Volunteer hierarchical structure	The tree like structure formed by new volunteers joining the system through the invitation of an existing volunteer.

1 Introduction

Two major challenges that our society faces are food wastage and food insecurity. Surplus food distribution has surfaced as a solution that can counter both these issues at once. With the increase in the numbers of existing and upcoming smart cities, this solution has become more scalable. These smart cities equip the movement with the use of Information and Communications Technology (ICT) and the Internet of Things (IoT). These provide excellent platforms that make the communication between the parties faster and enable the surplus to reach to a broader audience. Most importantly, the ICT and IoT based platforms provide an infrastructure that helps in the proper traceability and accountability of the surplus donations. This helps in providing greater transparency to the movement and better satisfaction to the involved parties.

In our previous work, we have provided an inaugural algorithm, Food Donor to Receiver Matching with Chronological Acceptance (FDRM-CA), that pairs donors with receivers of surplus food via willing volunteers [16]. It is noted that the availability of the volunteers fluctuates a lot and the success of the surplus movement empirically relies on their availability. Therefore, we have also provided an incentive module that goes hand in hand with the matching module stated above to boost and stabilize the availability of the volunteers who play a major role in the mobility of the surplus food [17].

While the existing incentive mechanism does attract more participation from the volunteers in addition to stabilizing their availability, it does not take into account the proactive behaviour of volunteers. It awards active and inactive volunteers alike, which may not be very encouraging for those who contribute proactively. In addition, volunteers may be encouraged to create an operational identity to invite probable proactive volunteers while they themselves can be passive participants only cashing in their share of the rewards. In this work, we modify the aforementioned incentive module so that it draws more participation from the volunteers and provides better incentives to the highly active participants while encouraging inactive participants to increase their contribution to the system. Our proposed mechanism achieves all the above mentioned enhancements while incurring the same cost to the incentive disbursal system.

Following is the organization of the paper. Section 2 presents the literature review in this field. Section 3 presents the problem formulation and the system structure along with the proposed mechanism. Section 4 provides the analysis of the same. Simulation results have been provided in Sect. 5 followed by Sect. 6 which presents the conclusion and the future scope for this work.

2 Related Work

Prior to this work, a foundation algorithm for matching food donors and receivers was provided [16], and an incentive mechanism was suggested for the volunteers who mediate this movement [17]. Herein, in this paper, we aim to provide an enhancement to the incentive mechanism so that the incentive system benefits further without incurring any extra volunteering cost overhead. That is to say, more active volunteer participation is encouraged by the modified incentive mechanism.

Food wastage is a major cause of concern for the whole world for the last few years hitting us hard on social and economic levels all around the globe [7]. Accumulating at an alarmingly high rate, the percentage of wasted food is now beyond 33% of the produce [18]. This itself amounts to billions of tons of food of which around a third is still perfectly edible but headed towards landfills [22]. There are several approaches to handle this serious concern. Amongst them, surplus food distribution has captured the fancy of the masses since it fights not only food wastage, but also food insecurity. Research is underway on the reinforcing effects of the Information and Communications Technology (ICT) on this long thriving notion. Design changes to the food supply chain have been forced by perishable food items [6]. With the application of the Internet of Things (IoT), routing and run time quality monitoring of the surplus has become a trend of the movement [11]. A steady increment in global popularity is observed for the surplus distribution platforms that are based on ICT [21]. The movement affects daily chores of people in terms of their interests and practices. On the other hand, organisations and cultures also have an impact on the movement itself making the people interact over food related topics [15].

Meanwhile, surplus distribution stands as a viable solution to the global climate change brought about by wasting food that is avoidable on the first place [12]. Volunteer participation and ICT mediation render these initiatives highly cost effective. A food collection body can be a great augmentation to the surplus distribution move by directly or indirectly sourcing, via Government mediated channels, the surplus [3]. A central surplus collection and distribution body using food distribution bags is the most economically viable solution. Hand in hand, ICT and IoT make a great pair in reducing the wastage of food and raising the awareness of the same among the masses [9]. A database of the current operational practices and methods of the market aids in the improvement of donor-receiver relations and leads to better suggestions on food habits [20].

Research with *OLIO* (https://olioex.com/), a surplus distribution app, reveals that policymakers may get encouraged in promoting massive application of ICT in the food industry owing the the fresh non-commercial relationships that form while interacting for surplus movement [8]. The promotion of surplus movement heavily relies on such prevailing legislative policies of the land [4, 5]. ICT currently exists in most food sharing community operations in the form of passive centralized communication platforms. This renders a broad scope of ICT usage in greater complexities that can aid in faster communication between interested parties, reach a broader audience and achieve better timings in food delivery while clocking better metrics in terms of tracking and scalability. An incentive mechanism delivered over an ICT based platform further increases and stabilizes the availability of volunteers making the movement of the surplus smoother and reachable to a broader audience, contributing further to the sustainability of the movement [17].

3 Algorithms and Methodology

With the support of technological platforms available in the existing and upcoming smart cities, surplus distribution has become a bigger phenomenon than ever. The addition of the incentive module only accelerates the same by promoting and stabilizing the availability of volunteers who take the major charge in the forward march of the movement. However, Active volunteers look forward to a greater reward from the system than the passive ones. This is addressed and discussed in this section.

3.1 Notations and Problem Formulation

Before we take up the aforementioned issue, we present the following notations that will help our problem representation.

- λ : Incentive amount given to a delivery person (volunteer). Total incentive disbursed for all volunteers taken together will be twice of this amount.
- n : Total count of volunteers taking the shortest path from the delivery person to the incentive module.
- κ : Total count of volunteers selected for incentive disbursal. The delivery person, who is always rewarded, is not included here.
- δ : Maximum allowable count of volunteers in the aforementioned chain of n volunteers to whom incentives will be disbursed.
- l_E : Percentage of most active volunteers from the chain of n volunteers who will be exploited for further surplus deliveries.
- d_E : Percentage of the final selection of k volunteers consisting solely of the aforementioned exploitable volunteers.
- Γ : Sequential list of the aforementioned n volunteers starting with the delivery person as the 0th volunteer.
- Δ : Sequential list of the activity counts of the n volunteers mentioned above.

While the matching algorithm matches the donors with the receivers, the incentive module ensures a consistent required availability of volunteers to mobilize the surplus. However, not all enrolled volunteers remain active in the system or participate equally in the activity. Therefore, the incentive module must differentiate, in terms of incentives, between those who proactively participate in the movement and those who do not. With a higher reward for the proactive volunteers, the system can be sure to retain their willingness to contribute.

3.1.1 The Proposed Incentive Module

We present the proposed Activity Aware Incentive Mechanism (AAIM) inspired by the multi-armed bandit mechanism [1, 13, 19], which rewards proactivity and penalizes inactivity. We use the settings similar to the one used in our previous work [17] based on the incentive scheme put forward in [2, 10, 14]. A representative of the same is provided in Fig. 1. Here, the volunteer γ delivers the surplus and makes the whole chain of volunteers, that is, α, β and γ eligible for incentives. However, β is an inactive person here and receives passive income. Whereas, α is a proactive participant and expects the system to recognize the extra effort in terms of an increased incentive.

Although, the issue seems simple and an over the top solution may be to increase the incentive of the proactive volunteer by some factor to keep the person interested in contributing more to the system, the effects of the same would be undesirable. Consider the step where only $\kappa = \lfloor\sqrt{n}\rfloor$ volunteers are rewarded. This is done to

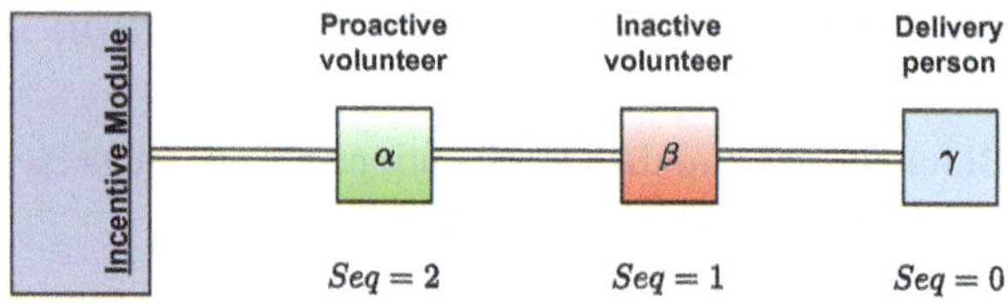

Fig. 1 Volunteer structure and proactivity for a successful delivery by volunteer γ

minimize the cost to the system. Therefore, increasing the incentive amounts of proactive volunteers would outright hurt the primary aim of cost minimization. On the other hand, the less active volunteers would definitely not be interested in contributing more if the system does not discriminate between the active and the passive volunteers. Thus, we propose a clever shift of funds from the passive volunteers to the active ones so as to keep the total cost to the system the same. This takes care of promoting the proactive nature of volunteers while encouraging the passive volunteers to be more active by penalizing them for a relatively lower activity count.

In Fig. 1, volunteer γ delivers the surplus. The volunteer hierarchical structure is formed as the volunteers invite other volunteers to the system who join through the inviter volunteer resulting in the parent-children construct. This parent-children construction of the volunteers is highlighted in Fig. 1. In this figure, for the surplus delivery, the chain of volunteers starts from the 0th volunteer γ, who actually delivers the surplus food. The chain consists of all the volunteers who are in the shortest path from γ to the volunteer α who is directly connected with the system. The chain consists of the volunteers γ, β and α, labelled with sequence numbers $Seq =$ 0, 1 and 2, respectively. These are the only volunteers considered for the disbursal of incentives for this surplus delivery completed by γ. Volunteer α is a proactive member of the chain while volunteer β is not. If selected, both these volunteers receive incentives using the formula $\Lambda_i = \$\frac{\lambda}{2^i}$ for a sequence number $Seq = i$. In the settings presented in Fig. 1, the existing mechanism will calculate incentives for γ as $\$\frac{\lambda}{2^0} = \$\lambda$, β as $\$\frac{\lambda}{2^1} = \$\frac{\lambda}{2}$, and α as $\$\frac{\lambda}{2^2} = \$\frac{\lambda}{4}$. The incentive amounts are directly proportional to the contribution of the respective volunteer in the movement of the surplus and reflects that γ has the maximum contribution in the same as the delivery person. Since, β has invited γ, β gets the next highest reward. Similarly, α gets the third highest reward for inviting β into the system. However, the existing mechanism does not select all n volunteers in the chain for reward. It randomly selects the κ volunteers to reward from the n volunteers in the chain. The proposed mechanism employs a different concept for selecting the κ volunteers for reward. Here, it utilizes the *exploit and explore* method from the concept of the multi-armed bandit [1, 13, 19], and modifies it to suit the settings of our requirements. In the multi-armed bandit concept of exploit and explore, the better performing resource is exploited more often than the one that performs less. To suit our requirements, we modify the same to further introduce a probabilistic selection of volunteers from the high performing group as shown in Fig. 2. Following are the steps for the selection:

1. Select the top $l_E\%$ volunteers with the highest activity. Here, $l_E\% < 50\%$ is a hyperparameter.

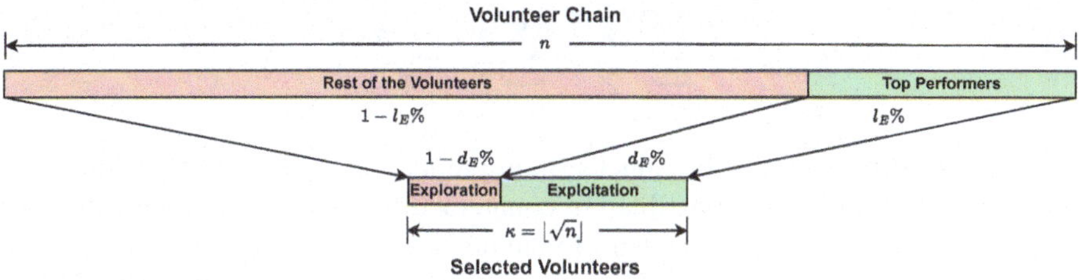

Fig. 2 Selection of $\kappa = \lfloor\sqrt{n}\rfloor$ volunteers from the chain of n eligible volunteers for incentive disbursal using the Exploit and Explore method

2. From the above selected list of volunteers, select at random $d_E\%$ of volunteers of the $\kappa = \lfloor\sqrt{n}\rfloor$ total volunteers to award incentives. Again, $d_E\% > 50\%$ is another hyperparameter.
3. Select the rest of the volunteers randomly from those volunteers having a lower activity to complete the selection of the κ volunteers.

Thus, we select the minority top contributors of the surplus mobility and exploit them further by selecting them more frequently (with higher probability) for providing incentives leading them to contribute more. Simultaneously, we select the less active volunteers less frequently for incentives, thereby effectively penalizing them for inactivity providing them with reason to increase their contribution. Note that the conditions $l_E\% < 50\%$ and $d_E\% > 50\%$ play an important role in the *exploit and explore* approach. It is this set of conditions that makes the exploitation part a majority as against the exploration part. Here, $l_E\%$ is the percentage of people who are the top performers of the volunteer team. We have taken $l_E\% < 50\%$, which indicates that we are flexible enough to fine tune it according to our need, and in the simulation we have varied it and have shown the impact. Similarly, $d_E\%$ dictates the percentage of people from the final selection who have been selected for exploitation from the aforementioned high performers. In our scheme, exploitation is the major percentage indicating that $d_E\% > 50\%$. We have also varied the same in the simulation and shown its impact. For the AAIM scheme, the incentive for a volunteer having the ith sequence number is given by the following.

$$\Lambda_i = \begin{cases} \lambda/2^i, & \text{if selected,} \\ 0, & \text{otherwise.} \end{cases} \quad (1)$$

With the increased probability of the high performers for being selected for receiving incentives, the expected total incentives received by these volunteers become more than that of them who were inactive. This way, proactivity is rewarded at the cost of penalizing inactivity. We now prove the financial feasibility of the AAIM scheme by providing an upper limit for the total incentive disbursed by the scheme for one surplus delivery.

$$\Lambda_{total} = \sum_{i=0}^{\kappa=\lfloor\sqrt{n}\rfloor} \frac{\lambda}{2^i} < \sum_{i=0}^{\infty} \frac{\lambda}{2^i} = \frac{\lambda}{1-\frac{1}{2}} = 2\lambda. \tag{2}$$

In the AAIM scheme, sybil attacks (multiple identities of a single volunteer to reap higher incentives) are possible. To counteract the same to a certain extent we adopt to limit the maximum number of volunteers who can receive incentives for a single surplus delivery to δ. This will motivate volunteers to keep identity duplication at a minimum so as to remain eligible for further incentives. Thus, in contrast to the existing mechanism, the proposed AAIM mechanism promotes proactivity and demotes inactivity while incurring no extra overall cost for the same.

3.2 Proposed Mechanism

The inclusion of an incentive module increases and stabilizes the availability of volunteers for the movement of surplus food. The AAIM scheme will act along with the matching scheme to encourage proactive behaviour of volunteers and discourage inactivity. Algorithm 1 presents the working of the scheme. It starts with initializing the incentives of all n volunteers in the volunteer chain of the delivery person. The number of volunteers to reward is calculated as $\kappa \leftarrow \min(\delta - 1, \lfloor\sqrt{n}\rfloor)$. It then divides the n volunteers into two groups. One is the *Exploit* group having $l_E\%$ of the topmost contributors in the mechanism. Random selection from this group contributes to $d_E\% \times \kappa$ of the selected volunteers. The other group is the rest of the volunteers from the n volunteers whose contribution did not help them qualify for the first Exploit group. This is the *Explore* group from which random selection makes up the rest of the κ selected volunteers. In addition, the delivery person (0th volunteer) always gets selected. Incentives are calculated for these $\kappa + 1$ people making that of the rest of the $n - \kappa - 1$ volunteers zero. This way, it connects with the multi-armed bandit mechanism where the best arms are selected for exploitation and from the rest exploration is performed. In this paper, the best arms signify the volunteers who are very active.

We start our incentive calculation from the volunteer who delivers the surplus food, and we denote this volunteer as the 0th volunteer. This is denoted like that, as the incentive formula (1) will be in conformity with the incentives we are paying to the volunteers. With this logic the 0th volunteer will get $\frac{\lambda}{2^0} = \lambda$ amount.

Algorithm 1: AAIM($\lambda, \delta, n, \Gamma, \Delta$)

$\Lambda \leftarrow \Phi$ // Initiate incentives.
$\kappa \leftarrow \min(\delta - 1, \lfloor\sqrt{n}\rfloor)$ // Count volunteers to reward.
$\Delta' \leftarrow$ **decreasing_sort**(Δ) // Sort by decreasing activity.
$\Delta'_{Cutoff} \leftarrow \Delta'[l_E\% \times n]$ // Get activity cutoff.
$V_{Exploit} \leftarrow \{t \mid (t \in \Gamma - \Gamma[0]) \cap \Delta[l] \geq \Delta'_{Cutoff}\}$ // Exploitation.
$V_{Explore} \leftarrow \Gamma - \Gamma[0] - V_{Exploit}$ // Exploration.
$\Gamma' \leftarrow \Gamma[0] \cup$ **random**($V_{Exploit}, d_E\% \times \kappa$)
$\cup$ **random**($V_{Explore}, (1 - d_E\%) \times \kappa$) // Assemble volunteers.
for $i \leftarrow 0$ **to** $n - 1$ **in** Γ **do** // For each volunteer.
  if $\Gamma[i] \in \Gamma'$ **then** // If selected.
    $\Lambda[i] = \lambda/2^i$ // Disburse incentive.
  else
    $\Lambda[i] = 0$ // No incentive.
return Λ // Return incentives.

4 Algorithm Analysis

Lemma 1 *The expected payment of the jth volunteer is bounded by* $\frac{2\lambda}{n}[\mu\nu + (1 - \mu)(1 - \nu)$.

Proof It is to be noted that with probability μ, a volunteer will belong to the top category and with $(1 - \mu)$ probability it will not. Again, we have that the volunteer may be selected either from the top category or from the non-top category. The probability of getting selected from the top category is ν and from the other category is $(1 - \nu)$. Now, the probability for the event A_1, that a volunteer will be selected is $\mu \cdot \nu + (1 - \mu)(1 - \nu)$. It is interesting to note that out of a chain of n volunteers, the selected volunteer maybe at any position in the chain. So, the probability that the volunteer will be selected at any i^{th} position can be calculated as:

$$\Pr(A_1 \cap A_2) = \Pr(A_1) \cdot \Pr(A_2) \tag{3}$$

where, A_2 is the event that the volunteer is from the ith position. It is to be noted that they are all independent giving $\Pr(A_2) = \frac{1}{n}$ as all are equally likely to be selected. We can write Eq. 3 as:

$$\Pr(A_1 \cap A_2) = \Pr(A_1) \cdot \Pr(A_2) = \frac{1}{n}[\mu\nu + (1 - \mu)(1 - \nu)]. \tag{4}$$

With this probability we can calculate the expected earnings of any arbitrary volunteer j as:

$$
\begin{aligned}
E[X] &= \sum_{i=1}^{n} \frac{\lambda}{2^i} \cdot \Pr(X = \text{the } i^{th} \text{ position for the selected case}). \\
&= \sum_{i=1}^{n} \frac{\lambda}{2^i} \cdot \frac{1}{n}[\mu\nu + (1-\mu)(1-\nu)] = \frac{[\mu\nu + (1-\mu)(1-\nu)]}{n} \cdot \lambda \cdot \sum_{i=1}^{n} \left(\frac{1}{2}\right)^i \\
&< \frac{[\mu\nu + (1-\mu)(1-\nu)]}{n} \cdot \lambda \cdot \sum_{i=1}^{\infty} \left(\frac{1}{2}\right)^i = \frac{[\mu\nu + (1-\mu)(1-\nu)]}{n} \cdot \lambda \cdot \frac{1}{1-\frac{1}{2}} \\
&= \frac{2\lambda}{n}[\mu\nu + (1-\mu)(1-\nu)]. \qquad (5)
\end{aligned}
$$

Observation 1 Plugging in the values of $\lambda = \$2000$, $n = 100$, $\mu = 0.25$ and $\nu = 0.75$ in Eq. 5, we get the expected maximum earning of an arbitrary volunteer as:

$$
\begin{aligned}
E[X] &= \frac{2\lambda}{n}[\mu\nu + (1-\mu)(1-\nu)] \\
\implies E[X] &= \frac{2 * 2000}{100}[0.25 * 0.75 + (1-0.25)(1-0.75)] \\
\implies E[X] &= \$15.
\end{aligned}
$$

5 Results

For the simulations, real data, sourced from OPEN DATA BARCELONA at https://opendata-ajuntament.barcelona.cat/data/en/dataset/est-cadastre-carrecs-us, was cast to the experiment settings using Python libraries. The dataset used for the simulation signifies cadastral positions according to the charge's use of the city of Barcelona. It has 22401 records. Each record states the district and the neighbourhood with their codes, the ownership description along with its valuation. The last column consisting of the value corresponding to the occupancy of land in some districts of Barcelona. So, in their data, each value corresponds to the valuation of a location. In our settings, we have used that value to correspond to a volunteer. We have cast the data like that way.

Volunteers were taken as 100 and successful surplus deliveries were varied in the range {2000, 4000, 6000, 8000, 10000}. Considering the number of volunteers available in a particular region, 100 is a sizable number. Similarly, the successful surplus deliveries were varied in the range {2000, 4000, 6000, 8000, 10000} as these are sizable and practical numbers. Our algorithm is scalable to experiment with higher numbers of volunteers and successful deliveries. The exploit group threshold was varied amongst $l_E\% = \{20, 40\}\%$. This was combined with the different values of selection of exploit group members for final incentive disbursal $d_E\% = \{60, 80\}\%$. The existing and the proposed methods were compared for the normalized incentives received by the volunteers when they are proactive versus when they are inactive.

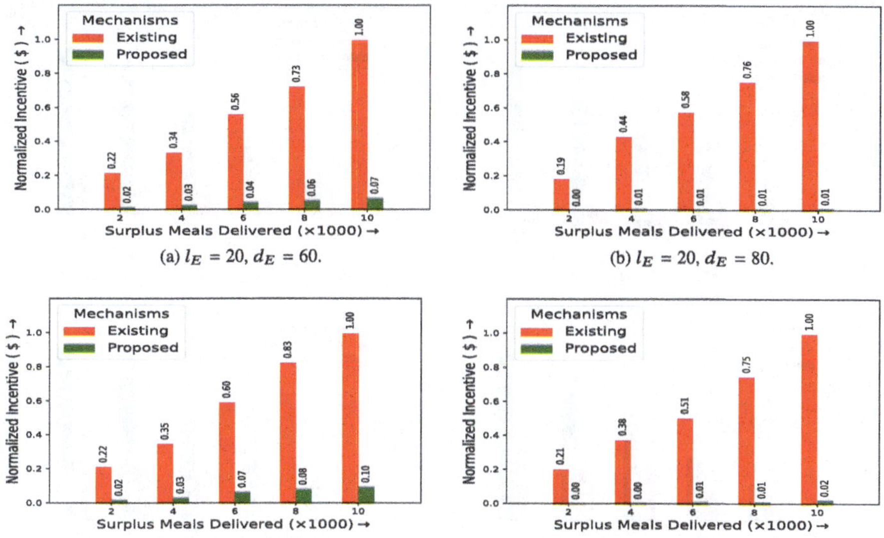

(a) $l_E = 20, d_E = 60$. (b) $l_E = 20, d_E = 80$.

(c) $l_E = 40, d_E = 60$. (d) $l_E = 40, d_E = 80$.

Fig. 3 The AAIM incentive scheme for *inactive* volunteers while varying $l_E\%$ and $d_E\%$

The total incentive disbursed by the AAIM method for these two types of volunteers were also compared with that of the existing method to deduce the proactivity bonus and the inactivity penalty amounts by the proposed mechanism.

5.1 Analysis

The results presented in Figs. 3, 4 and 5 are analysed here. Figure 3 shows the reduction in incentive of the inactive volunteers by the AAIM scheme when compared with the existing scheme for different values of exploitation threshold $l_E\%$ and selection from exploit group for incentive disbursal $d_E\%$. It shows that the proposed mechanism penalizes inactivity heavily. On the other hand, Fig. 4 shows the bonus that the proactive volunteers receive using the AAIM mechanism in contrast to that for the existing mechanism while varying $l_E\%$ and $d_E\%$. This shows that the proposed mechanism encourages the proactivity of the volunteers.

Lastly, Fig. 5 contrasts the activity based rewarding nature of the AAIM scheme against the flat rewarding nature of the existing scheme. For the inactive volunteers, the reduction in incentive by the proposed mechanism against the existing mechanism is equivalent to a penalty for inactivity. Similarly, the increased incentive disbursed by the proposed mechanism for the proactive volunteers when contrasted with the incentive rewarded by the existing mechanism is equivalent to a bonus for proactivity. Both these activity based penalty and bonus systems are enforced without incurring

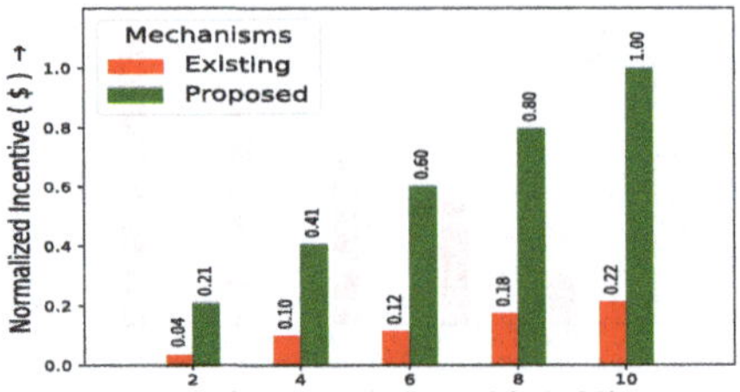

(a) $l_E = 20, d_E = 60.$

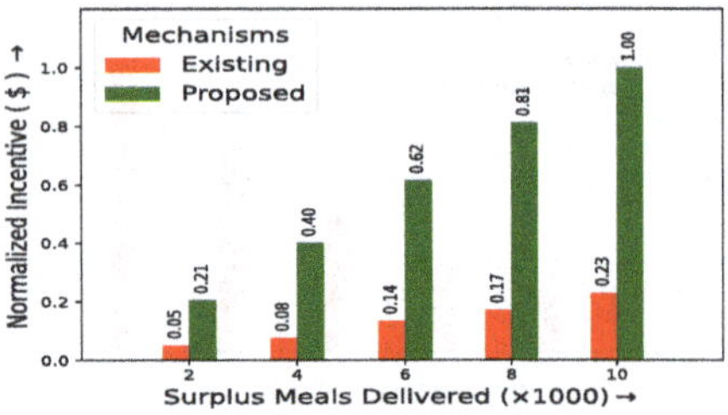

(b) $l_E = 20, d_E = 80.$

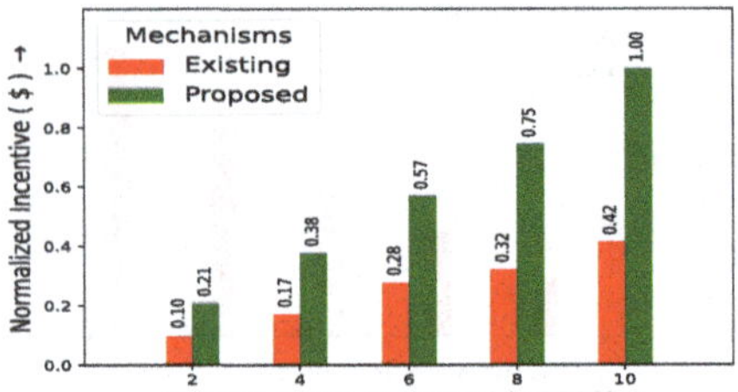

(c) $l_E = 40, d_E = 60.$

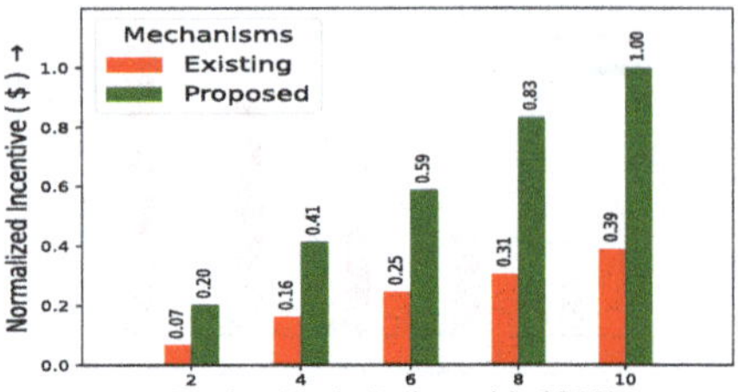

(d) $l_E = 40, d_E = 80.$

Fig. 4 The AAIM incentive scheme for *proactive* volunteers while varying $l_E\%$ and $d_E\%$

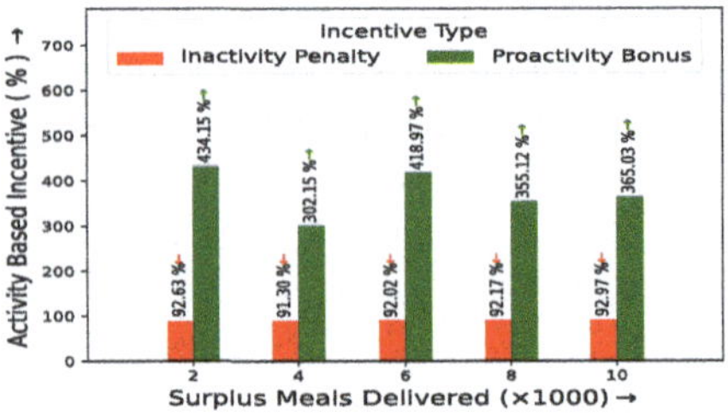

(a) $l_E = 20, d_E = 60.$

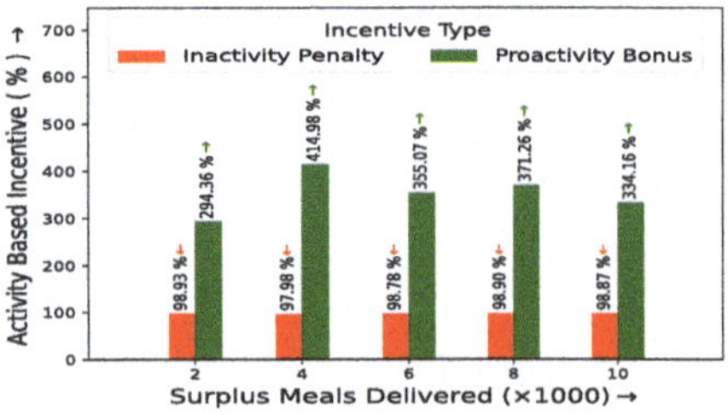

(b) $l_E = 20, d_E = 80.$

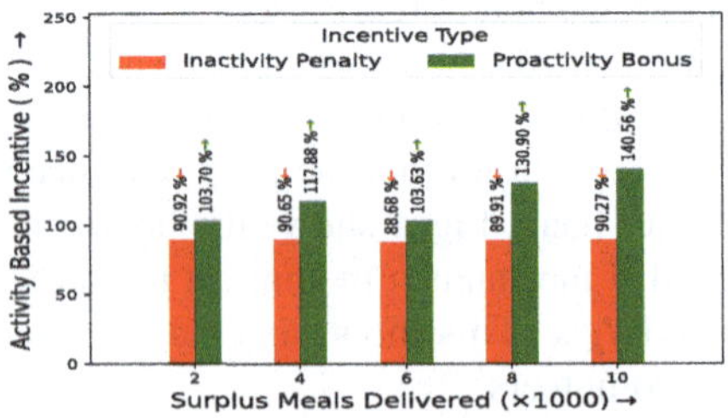

(c) $l_E = 40, d_E = 60.$

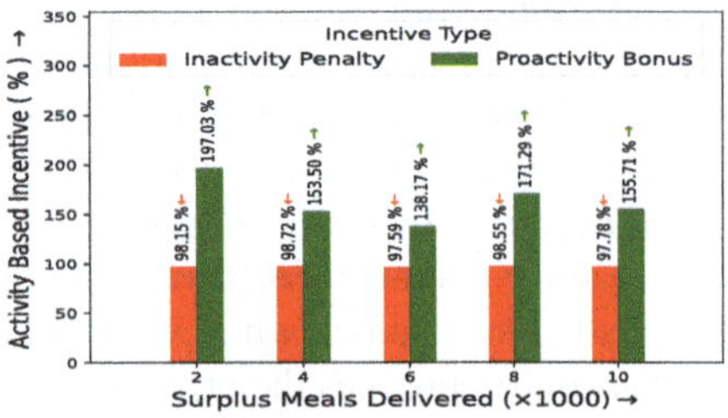

(d) $l_E = 40, d_E = 80.$

Fig. 5 The AAIM scheme bonuses and penalties for volunteers while varying $l_E\%$ and $d_E\%$

any extra cost to the incentive module, i.e., both the existing and the proposed schemes disburse the same total incentives when all volunteers are taken together.

6 Conclusion

Food wastage is a great challenge for the world, as is food insecurity. Surplus food distribution has come up as a great solution for both the challenges until deeper rectifications are able to transform the situation. Volunteer participation plays a pivotal role in the success of the surplus movement. Therefore, an incentive mechanism can not only stabilize their availability, but also encourage a greater participation. Existing incentive mechanisms provide for a basic start to this goal. In real life environments, all volunteers do not contribute equally to the movement and those who contribute more expect to be rewarded more for their proactive participation. However, the existing mechanisms do not account for the degree of contribution of the volunteers and reward all volunteers flatly with respect to the same.

The proposed incentive scheme uses the concept of the "multi-armed bandit problem" to take into account the degree of participation of the volunteers while considering for incentive disbursal. For inactive volunteers it reduces the incentive amount disbursed to them. This amount that was saved from the incentives for the inactive volunteers is transferred to the proactive volunteers. Thus, the proposed incentive mechanism discourages inactivity and promotes proactivity while incurring the same cumulative incentive amount disbursed when compared with that of the existing mechanism.

We note that the proposed incentive scheme is vulnerable to sybil attacks (identity duplication). Limiting the maximum possible number of volunteers to reward as a solution to the same may be counter productive towards encouraging the volunteers to join the movement through invitations. Designing and pruning a mechanism to address the sybil attacks is a challenging future work.

Acknowledgements Fatos Xhafa's work is supported by Project "Optimizing Carsharing and Ridesharing Mobility in Smart Sustainable Cities" (grant 21S09355-001) of the Ajuntament of Barcelona and the Fundació "la Caixa" under the framework of the Barcelona Science Plan 2020-2023.

References

1. Auer P, Cesa-Bianchi N, Fischer P (2002) Finite-time analysis of the multiarmed bandit problem. Mach Learn 47(2):235–256
2. Babaioff M, Dobzinski S, Oren S, Zohar A (2012) On bitcoin and red balloons. In: Proceedings of the 13th ACM conference on electronic commerce, pp 56–73
3. Bergström P, Malefors C, Strid I, Hanssen OJ, Eriksson M (2020) Sustainability assessment of food redistribution initiatives in Sweden. Resources 9(3):27

4. Davies AR, Cretella A, Franck V (2019) Food sharing initiatives and food democracy: practice and policy in three European cities. Polit Govern 7(4):8–20
5. Eriksson M, Giovannini S, Ghosh RK (2020) Is there a need for greater integration and shift in policy to tackle food waste? insights from a review of European union legislations. SN Appl Sci 2(8):1–13
6. Esteso A, Alemany M, Ortiz Á (2021) Impact of product perishability on agri-food supply chains design. Appl Math Modell
7. Ganglbauer E, Fitzpatrick G, Molzer G (2012) Creating visibility: understanding the design space for food waste. In Proceedings of the 11th International conference on mobile and ubiquitous multimedia, pp 1–10
8. Harvey J, Smith A, Goulding J, Illodo IB (2020) Food sharing, redistribution, and waste reduction via mobile applications: a social network analysis. Ind Mark Manage 88:437–448
9. Koivunen L, Laato S, Rauti S, Naskali J, Nissilä P, Ojansivu P, Mäkilä T, Norrdal M (2020) Increasing customer awareness on food waste at university cafeteria with a sensor-based intelligent self-serve lunch line. In: 2020 IEEE international conference on engineering, technology and innovation (ICE/ITMC). IEEE, pp 1–9
10. Messner M, Pavoni N, Sleet C (2012) Recursive methods for incentive problems. Rev Econ Dyn 15(4):501–525
11. Nagarajan SM, Deverajan GG, Chatterjee P, Alnumay W, Muthukumaran V (2022) Integration of IoT based routing process for food supply chain management in sustainable smart cities. Sustain Urban Areas 76:103448
12. Nikravech M, Kwan V, Dobernig K, Wilhelm-Rechmann A, Langen N (2020) Limiting food waste via grassroots initiatives as a potential for climate change mitigation: a systematic review. Environ Res Lett
13. Padmanabhan D, Bhat S, Prabuchandran K, Shevade S, Narahari Y (2022) Dominant strategy truthful, deterministic multi-armed bandit mechanisms with logarithmic regret for sponsored search auctions. Appl Intell 52(3):3209–3226
14. Roughgarden T (2016) Cs269i: incentives in computer science lecture# 10: Incentives in crowdsourcing
15. Rut M, Davies AR, Ng H (2020) Participating in food waste transitions: exploring surplus food redistribution in Singapore through the ecologies of participation framework. J Environ Policy & Plann 1–14
16. Sanyal S, Kumar Singh V, Xhafa F, Sanyal B, Mukhopadhyay S (2021) A game theoretic framework for surplus food distribution in smart cities and beyond. Appl Sci 11(11):5058
17. Sanyal S, Mukhopadhyay S, Xhafa F, Singh VK, Sanyal B (2022) An incentive framework for surplus food distribution in the context of smart cities. In: International conference on network-based information systems. Springer, pp 35–47
18. Shi ZR, Yuan Y, Lo K, Lizarondo L, Fang F (2020) Improving efficiency of volunteer-based food rescue operations. In AAAI, pp 13369–13375
19. Slivkins A et al (2019) Introduction to multi-armed bandits. Found Trends® Mach Learn 12(1-2):1–286
20. Spicer LC, Patel D, Bodurtha ER, Maslen AM (2020) Surplus food management: a case study. Student Work
21. Spring CA, Biddulph R (2020) Capturing waste or capturing innovation? comparing self-organising potentials of surplus food redistribution initiatives to prevent food waste. Sustainability 12(10):4252
22. Zhong R, Xu X, Wang L (2017) Food supply chain management: systems, implementations, and future research. Ind Manag Data Syst

Evaluation of Smart Charging Integrated with Smart Energy Management and Advance Booking in an eMobility Urban Living Lab

Shanshan Jiang, Marit Natvig, Svein Hallsteinsen, and Karen Byskov Lindberg

Abstract The wide adoption of electric vehicles (EV) is crucial for the decarbonization of transportation and the vision of smart cities. Solutions for smart EV charging and energy management have been proposed, but there are few reports on realization and experience of adopting them in real life. The GreenCharge project has proposed, demonstrated, and evaluated digital support for smart and green charging in an urban living lab setting. The proposed solutions are documented in a reference architecture meant to act as a blueprint both facilitating the extension and integration of the involved systems in the prototype implementation and supporting replication. However, successful uptake also depends on the implementation process, among others, successful motivation and engagement of relevant stakeholders. Thus, the evaluation addressed in this chapter is based on the CIVITAS process evaluation methodology carried out within the context of design science and urban living labs. We share our experience and lessons learned from the design, implementation, and deployment of the proposed solutions in an urban housing cooperative in Norway, focusing on smart charging integrated with smart energy management and advance booking of charge points. Barriers and drivers regarding this innovation process are identified and recommendations to overcome the barriers are suggested. Stakeholder acceptance evaluation of the implemented solutions are presented as well. The findings are intended to help stakeholders and policy makers to develop successful strategies for sustainable electric mobility and electric energy supply.

S. Jiang (✉) · M. Natvig · S. Hallsteinsen
SINTEF Digital, Trondheim, Norway
e-mail: Shanshan.Jiang@sintef.no

M. Natvig
e-mail: Marit.K.Natvig@sintef.no

S. Hallsteinsen
e-mail: Svein.Hallsteinsen@sintef.no

K. B. Lindberg
NTNU Electric Power Engineering, Trondheim, Norway
e-mail: karen.lindberg@ntnu.no

SINTEF Community, Oslo, Norway

R. Shinkuma et al. (eds.), *Advances in Engineering and Information Science Toward Smart City and Beyond*, Engineering Cyber-Physical Systems and Critical Infrastructures 5, https://doi.org/10.1007/978-3-031-29301-6_5

Acronyms

CP	Charge point.
CPO	Charge Point Operator.
DSO	Distributed System Operator.
eMobility	Electric mobility.
EMP	eMobility Provider.
ESN	Energy smart neighborhood.
EV	Electric vehicle.
LEM	Local Energy Manager.
LMS	Load Management System.
PV	Photovoltaic.
RA	Reference architecture.
RES	Renewable energy source.
SEM	Smart energy management.
SoC	State-of-charge.
ULL	Urban living lab.
UML	Universal Modelling Language.
V2G	Vehicle-to-grid.

Glossary

Advance booking	Means that an EV user can reserve a CP for a time slot starting in the future and also provide information about energy demand.
Demand flexibility	Means to which extent a predicted energy demand can be shifted in time or power. For example, if you start a dishwasher in the morning, you may not need the clean dishes until you return from work in the evening, or when you connect an electric vehicle for charging in the afternoon, you may not need the vehicle until the next morning, or the water boiler needs not be as hot in periods of low demand for hot water, so the power can be temporarily reduced or switched off. GreenCharge has a particular focus on leveraging the demand flexibility of charging electric vehicles.
Distributed System Operator (DSO)	Responsible for operating and maintaining the electricity distribution grid.
Energy smart neighborhood (ESN)	A microgrid composed of smart buildings, charging stations and other energy consumers and producers that use an ICT infrastructure

	and a centralized or distributed energy management systems to optimize energy usage.
Flexible charging	Means that the charging can be done at any time before a deadline. In such cases, the charging can be distributed over time and adapted to the availability of energy and transfer capacity according to predefined rules (e.g., that the use of green energy should be prioritized).
Process evaluation	Evaluation of the processes of preparation, implementation, and operation of measures, including the roles of information, communication, and participation.
Reference architecture	Means that the focus of the architecture is on the integration of logical services, from different providers, into an ecosystem for smart and green charging, instead of on an architecture for a concrete system implementation.
Smart charging	Refers to charging that exploits demand flexibility to balance the demand and the production of electricity by means of information and communication systems. It is often supervised by an external control system.
Smart energy management (SEM)	System functionality for optimising the use of energy adapted to energy demands and availability. May also take predictions of future energy demand and availability into account.
State-of-charge (SoC)	An indication of the amount of energy stored in a battery. It is given as a percentage, meaning the percentage of the full capacity currently available in the battery. The SoC is difficult to measure accurately, but several methods are available to give an approximate value, and most electric vehicles have an instrument on the dashboard showing the SoC.
Vehicle-to-grid (V2G)	The capability of an electric vehicle to behave as a stationary battery, returning accumulated energy stored in the vehicle battery to the grid.

1 Introduction

The electrification of transportation is an important driver for the vision of smart cities. This is because electric mobility (eMobility) provided by electric vehicles (EV) partially powered by local renewable energy sources (RES) is key to the decarbonization of the transport sector [5]. The adoption of EVs is not simply about encouraging people to buy EVs, but also about overcoming a number of other barriers. Firstly, charge points (CP) are not widespread yet and some of them are dedicated to their subscribed users. In areas where the EV density is high, such as in Norway, EV drivers may have to queue up and wait for access to chargers. Potential EV users may thus hesitate to buy EVs as they worry about where their EVs can be charged when needed (so-called *charge anxiety*). Secondly, building new charging infrastructure is a huge investment and charging infrastructure providers may be concerned about getting enough users and how the CPs can be best utilized to get a good return of investment. Finally, EV charging creates extra electric energy demand, and charging a large number of EVs simultaneously in a neighborhood may create power peaks possibly exhausting the capacity of the existing local electricity infrastructure. On the other hand, EV charging also offers considerable demand flexibility. A study on home charging found that EVs usually complete charging within 3 h after plugged-in, but remain connected for 12 h on average for overnight charging [21].

Smart charging addresses these barriers by means of information and communication technologies. It exploits demand flexibility to balance the demand and the production of electricity, and thus reducing the capacity problems in the electricity supply grid. Furthermore, it offers booking of charge points to ensure predictable access and avoid queues, but also to predict charge demand and demand flexibility. Thus smart charging is crucial both to the sustainability and the acceptance of eMobility.

Today EV users may get an online overview of the location of charge stations and the availability of CPs, but although various solutions have been proposed, support for booking is still very seldom. One reason is that CP owners fear that booking may reduce the utilization of their equipment and their return of investment.

Booking solutions vary in the degree of charge planning support and to which extent a reservation can be guaranteed [2]. The *reserve now* scheme means that the EV user can reserve a CP which is currently free, and it will be blocked for use by others until the user releases it. This is already supported by commonly used protocols in the field and offered by some charge stations. Usually, the user can only reserve shortly before arrival to the charge station, and if the user has not arrived within the time limit, the reservation is cancelled. Thus the support for charge planning is rather poor.

Advance booking is a more sophisticated scheme where the EV user can reserve a CP for a time slot starting in the future, and usually also requires the user to provide information about energy demand. This provides better support for charge planning for the EV user. It also enables the use of the CP by others in the period between the placement of the reservation and the start of the reserved period. Better charge planning also means better prediction of electric energy demand from EV charging, which is key to optimize the utilization of both the available charging equipment and the electricity supply system.

Traditional subscription-based business model and access to CPs is another barrier to wide adoption of eMobility. Roaming with credit card-based payment is a promising solution to this barrier. Similar to the roaming of mobile services, the roaming of EV charging services allows EVs to be charged at any CP regardless of who is operating the facility.

The GreenCharge project (https://www.greencharge2020.eu/) has proposed and tested innovative solutions to help address the above challenges [10]. The GreenCharge solutions for smart and green charging are based on energy smart neighborhoods (ESN) that offer *predictive and optimal energy management* of the whole neighborhood. In such an ESN, EV charging is coordinated with other local energy demand, local production and capacity constraints in the public electric energy supply. Demand flexibility (from EV charging and other energy demands in the neighborhood such as heating and cooling devices) and local energy storage capacity are exploited, taking predictions of current and future energy demands and the current and predicted energy production from local RES into account. In addition, solutions for flexible sharing of the charging infrastructure and flexible sharing of various kinds of EVs are proposed. Moreover, the GreenCharge concept also includes viable business and price models for eMobility adoption, for example, rewarding charging behavior contributing to peak reductions and encouraging less consumption and more sharing (including sharing of EVs and CPs).

The methodology adopted in GreenCharge is a design science [8]-based iterative innovation process combined with the urban living lab (ULL) concept. To facilitate learning and future replication, GreenCharge has demonstrated how technological solutions and associated business models can be integrated and deployed to overcome the barriers to massive eMobility adoption through seven demonstrators in ULLs in Oslo (Norway), Bremen (Germany) and Barcelona (Spain). Each demonstrator implements a number of eMobility measures [10]. These measures cover charging, energy management and business aspects and support the realization of the innovative solutions. Technology prototypes for the demos are based on integration of commercial systems owned by project participants. To facilitate the integration of such systems in a well-defined manner, GreenCharge has defined a reference architecture (RA) for a full-fledged specification of the ecosystem describing the concept and the required functionality, interfaces and collaborations as well as sharing of responsibilities between the participating systems [12, 13].

Innovative solutions for smart charging and booking of CPs have been proposed in the literature (see related work in Sect. 2), but there are few reports on realizations and experience of adopting them in real life. This chapter aims to fill the gap and report the experience from real-life demonstration of innovative eMobility solutions in Oslo. The Oslo demonstration addressed the needs of a typical housing cooperative with legacy electricity infrastructure, where poor access to home charging for people living in urban building blocks is hindering EV adoption in urban areas. Two demonstrators have been set up focusing on smart energy management utilizing local RES for charging (Demo 1) and flexible sharing of private charging facilities through advance booking and roaming (Demo 2).

The real-life demos in Oslo featured a highly interdisciplinary approach to cope with the complexity in the implementation of innovative eMobility measures in practice. This chapter presents the design, implementation and deployment of the demos and the lessons learned from en evaluation of the whole implementation process (from the design to the demonstration lasting from October 2018 to February 2022). Through the analysis of the technical, social, and organizational challenges encountered, recommendations for eMobility implementation strategies are provided to facilitate future replication. The related research questions are:

- RQ1: How was the stakeholder acceptance of the innovative measures?
- RQ2: Which drivers and barriers affected the work?
- RQ3: How can the implementation process of smart and green charging be supported?

This chapter is an extended version of [11], which provided a description of the process evaluation of Oslo Demo 1. The main extensions in this chapter include (1) a description of the design, implementation and process evaluation of Oslo Demo 2, (2) a description of the business aspect measures of Demo 1, (3) an evaluation of the stakeholder acceptance for the implemented measures in the two demos, (4) an description of the key supporting activities for the demos, and (5) a more elaborated description of methodology. The main contributions of the chapter are:

- A description of the design and implementation of the two demos in Oslo ULL;
- Stakeholder awareness and acceptance evaluation of the innovative eMobility and energy management measures implemented in the demos;
- A description of drivers, barriers and lessons learned from the implementation of the demos and recommendations for future eMobility implementation strategies.

Norway is a world leader in EV adoption, and people have high demands and willingness to invest in EVs (new EVs constituted 65% share of new car sales in 2021[1]). Thus, the evolution and deployment of eMobility in the Oslo demos are particularly suited to gather valuable experience and may be representative for future evolution of eMobility in other countries.

In the following, Sect. 2 discusses related work and Sect. 3 describes the methodology, including the approach for the overall innovation process and for the evaluation and learning. Section 4 presents the Reference Architecture, while Sect. 5 describes the planning and implementation of the two demos, including the measures implemented and the supporting prototypes. Stakeholder acceptance evaluation and lessons learned from the process evaluation are presented in Sect. 6. Section 7 presents the recommendations derived from the process evaluation and discusses the deployment of smart charging in large urban areas and the validity of the results. Finally, Sect. 8 concludes the chapter.

[1] https://elbil.no/content/uploads/2022/05/Arsberetning-2021-for-Norsk-elbilforening.pdf.

2 Related Work

ENTSO-E has provided a deep analysis of eMobility and its impact on the power system [5], focusing on interfaces between transport and energy sectors. The GreenCharge concept and the Reference Architecture are consistent with the framework and recommendations proposed there, but adds interaction with the building sector. The idea is to optimize use of local RES and reduce power peaks by coordinating EV charging demand with other electric energy demand and production from local RES in the neighborhood where the charging facilities are located, exploiting demand flexibility on both sides.

Many existing technology and services that facilitate charging of a group of EVs in a parking place involve a *reactive* management system, which ensures that the aggregated electricity load for charging several EVs does not surpass the available capacity in the parking place. In this case, all EVs immediately start charging after plugging in. If the capacity limit is reached, all EVs will have their charging power reduced, either by reducing the charging speed or charging the cars randomly in sequence. Hence, the reactive management system provides a safe solution for charging of multiple EVs, but the EV users might become dissatisfied if the desired charging demand is not met. Typically management of a number of CPs in a common garage is equipped with such simple load balancing systems.[2] Another early technical solution manages the CPs by simply disconnecting/reconnecting the chargers when problems occur/are solved as detected in the low-voltage network, following a hierarchical corrective approach [19].

Smart charging refers to charging supervised by an external control system, as opposed to the above reactive management-based charging. Demand management via smart charging is explored in [3], where the charging current of an EV charger can be varied using an external signal, for example, to reduce the charging speed when the exhaustion of the energy supply capacity is approaching. This approach was demonstrated in a two-year field trial involving about 700 drivers of a variety of EV models. The findings from the trial show that high flexibility levels are available from home charging and smart charging can exploit this flexibility to reduce peak demand due to EV charging [3].

The GreenCharge approach to smart charging exploits the flexibility in EV charging as well, and is based on a *predictive* smart energy management (SEM) system with optimal scheduling. The SEM system provides an optimal plan for scheduling the charging of individual EVs based on the user's expected driving demand. The optimal plan is calculated based on the departure time of each EV, and its charging need (the difference between the actual state-of-charge (SoC) at plugging-in, and the desired SoC when plugging out). In such a system, the charging of EVs that will be unplugged the next day can be postponed (temporarily stopped), and EVs that plan to

[2] Example commercial solutions like Zaptec (https://zaptec.com/elbillading-borettslag-sameie/), EVlink Load Management System from Schneider Electric (https://www.se.com/ww/en/product-range/62159-ecostruxure-ev-charging-expert/), Ohmia Charging (https://www.ohmiacharging.no/elbillading-i-borettslag-eller-sameie/).

leave within the next few hours can be prioritized. The GreenCharge approach allows the control of the start and stop time of the chargers, in addition to the charging speed. While [3] focuses on controlling individual chargers, GreenCharge addresses both the control of individual chargers and the aggregation of all the charging demands and other energy demands in a building/neighborhood.

In the literature, optimization models are used in a simulation environment to investigate predictive smart energy management [9, 18, 20]. However, only a few demos have been tested in real life, and the experiences are not always successful. For instance, in the INVADE project (https://h2020invade.eu/the-project/), the aim was to design and develop a flexibility management system based on optimization. However, the demos in the project experienced large challenges in the transfer of real-time data between the different stakeholders, and the implementation of the optimal scheduler [17].

There exist large scale trials of real-life EV charging in various contexts. The My Electric Avenue project performed a large scale EV trial in the United Kingdom with over 200 Nissan LEAFs (with a battery capacity of 24 kWh), where the EV onboard monitoring system automatically collected charging data, such as the charging start time and the initial and final SoC. Lessons learned from the trial regarding the user charging behavior, EVs' impact on the low-voltage networks, and the effectiveness of the proposed EV management solution are presented in [19]. Another related work to residential EV charging pilot study is presented in [1], which shows that flexibility-differentiated pricing and control can be effective in minimizing residential EV charging impact on peak electricity demand through reshaping EV charging loads. GreenCharge offers similar monetary incentives and predictive control. In addition, GreenCharge smart charging is coordinated with local photovoltaic (PV) production and other energy demands in the neighborhood, which [1, 19] do not address.

Regarding advance booking of CPs, there are a few papers proposing solutions, but very few realizations. Flocea et al. [6], Orcioni and Conti [16] both propose extensions to the OCPP protocol to enable advance booking and report on prototype implementations, but no real-life trials. Hardinghaus et al. [7] reports on a pilot project in the city of Hamburg, where public charging facilities in a densely populated urban neighborhood with limited parking were equipped with a reservation option providing exclusive access for a group of local residents and businesses. Several Apps provide an overview of publicly available charging points including type of chargers and charging capacity (kW). However, none of them offers booking services prior to arrival. The Zap-Map for UK offers an overview of both private and publicly owned CPs in a map. For privately owned home chargers, an agreement for charging can be made by a chat-function where the price and method of payment are settled bilaterally.[3] In Norway, a few commercial providers[4] offer payment solutions for sharing private CPs with others, but with no reservation solutions. A pilot research project "Spot-on" has developed a platform where people can make their privately

[3] https://www.zap-map.com/.

[4] Such as Cloudcharge, Charge 365, Smartcharge and Plugpay.

owned CPs available for others, however, it is still in the early phase and is not yet launched in the market[5] In comparison, the GreenCharge solution is a more general one, based on extensions to the roaming protocols in order to enable seamless booking and access to all across the emerging plethora of operators and service providers in the EV charging business. The App offers a calendar based user interface, avoids double bookings, and offers a safe credit-card payment solution.

3 Methodology

Urban living labs (ULL) have been a popular approach, especially in the context of smart cities, to addressing the challenges related to sustainable urban interventions. Steen and van Bueren [22] summarized the characteristics of urban living labs as follows: (a) *aimed* at innovation, formal learning for replication and increasing urban sustainability; (b) covering *activities* of development (all phases of the product and service development process), co-creation and iteration (feedback, evaluation and improvement); (c) *participants* from public and private sectors, users and knowledge institutes, all with decision-making power; (d) innovations taking place in the *real-life use context*. They further identified five overarching phases in the innovation process: research, development, testing, implementation and commercialization.

Our methodology is a combination of ULL and design science [8]. This approach supports the overall iterative innovation process to realize an innovative charging infrastructure as envisioned in the GreenCharge concept, as illustrated in Fig. 1. The demonstration and evaluation have been performed in real-life settings, and a multidisciplinary project team, residents and the housing cooperative administration have been actively participated in the co-creation of the innovative solutions.

As illustrated in Fig. 1, the concept elaboration and design leads to the design and specification of the Reference Architecture (RA). This is based on requirements

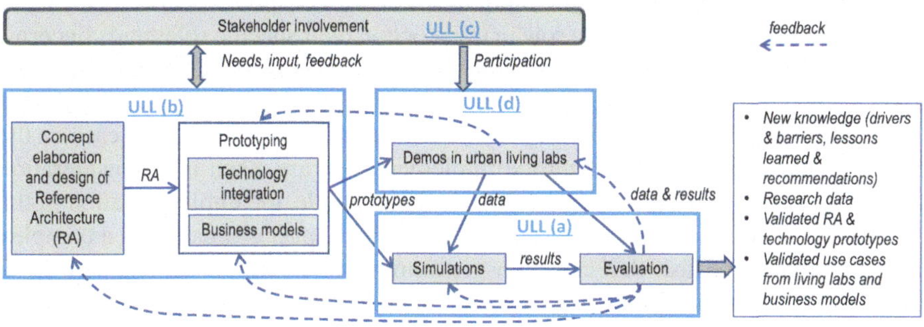

Fig. 1 The overall innovation process

[5] See https://its-norway.no/spot-on-bedre-ladelosninger-gjennom-pilot-t/ and https://atlas.powerzeek.com/.

derived from the needs and use cases of the demonstrators, and on the knowledge of and insight from existing architectures, technology and infrastructure. Existing results, such as standards and standardized interfaces are reused whenever possible. The technology prototypes integrating software and hardware are built as a result of new developments, refinements, extensions and integration of commercial systems owned by project participants, where the integration architecture and interfaces are in line with and supported by the RA. The business models with appropriate price models are developed through participatory design and workshops with relevant stakeholders. The close and active involvement of stakeholders (such as residents, housing cooperative administration, EV users, charge point operators, eMobility providers and policy makers) ensures that the requirements (from user needs, policies, standards, etc.) are met and that the prototypes and designs are relevant/correct. To evaluate the solutions, the prototypes with supporting business models are tested in real-life demonstrators in ULLs. Feedback from the evaluation and stakeholder involvement also facilitates the refinement of the RA, the technology prototypes and business models. Research data are collected from the demonstrators and used as input to simulations to complement the real-life demonstrators. New knowledge are obtained through the evaluation results and lessons learnt from the demonstrations and simulations.

The GreenCharge evaluation builds on the CIVITAS evaluation framework [4] and consists of *impact evaluation* (to investigate the effects of the measures utilizing an indicator framework) and *process evaluation* (to learn about the innovation process through collation and analysis of activities throughout the whole process). The principles for the impact evaluation are described in [4]. Details on the approach and indicators used in the impact evaluation are described in [14]. This chapter focuses on the user acceptance and awareness aspects of the impact evaluation and the process evaluation covering the design, implementation and operational stages of the Oslo demonstrators' implementation process.

The process evaluation is based on qualitative analysis and focuses on identifying drivers, barriers and risks, as well as required supporting activities to help understand why measures succeed or fail. The input to the analysis include: (1) *Pre-analysis* of stakeholders having a significant role in the process, their specific roles and concerns, as well as implementation risks, possible barriers and drivers. (2) *Monitoring and assessment* of relevant actions and events to understand what has happened and why. The monitoring identifies supporting activities that have a significant influence on the process. Other activities affecting the implementation are taken into consideration even though they may seem to be irrelevant. For example, the implementation of the automatic collection of research data needed in the impact evaluation is a complex and comprehensive task that caused a great number of discussions and actions in parallel to the implementation of the measures. (3) *Stakeholder involvement* to collect input on stakeholders' experience and learning through meeting minutes and logs documenting challenges, events and decisions during this process. (4) *A focus group* arranged with all actors involved in the ULL to investigate the barriers and drivers encountered and the effect of supporting activities, where a neutral facilitator asked open questions and the participants discussed. The input has been analysed and the results are summarized in Sect. 6.

4 Reference Architecture (RA)

The GreenCharge innovations require cross sectorial collaboration involving business actors and supporting technical systems of the energy supply, transport and building sectors. The electric energy supply and the building sectors are mature and highly regulated sectors with well-established business structures and supporting technical systems. In the younger eMobility sector, a business structure with business actor roles, supporting appliances as well as technical and business systems has already emerged as well. Therefore, our approach to realizing the GreenCharge concept is to extend the functionality of and the collaboration between these already existing systems.

In line with this approach, the RA aims to specify the participation of relevant existing systems in the realization of the GreenCharge solution in terms of modified and/or added responsibilities and collaboration patterns necessary to support the GreenCharge concept. This is described in terms of Universal Modelling Language (UML) stakeholder-, use case-, decomposition-, and collaboration models in the RA document [13]. An initial version was developed in the beginning of the project in co-design between the research and commercial partners to facilitate a common understanding of the GreenCharge concept and serve as a blueprint for the implementation of the digital support for prototypes used in ULLs. Then it has been refined towards the end of the project to reflect lessons learned during the implementation and use of the prototypes in ULLs.

Figure 2 gives a high-level view of the use cases (ovals) with stakeholder roles (stickmen), important technical equipment (pictograms), and software services supporting them (rectangles). The key elements and the extension needed to support the GreenCharge innovative measures are provided by the services: *Charge Service Provisioning* supports eMobility Providers (EMP) in the selling of charging services to EV drivers. This involves authorizing the use of selected CPs and taking care of billing and payment. The user support is provided by the *Charge Planning Assistance* service assisting EV drivers in planning the charging of their EVs and booking charging at suitable CPs. This is an **innovative service not commonly provided** in existing charging stations. It is to be provided by EMPs as an extension of their offer to their customers (EV drivers) and provides important information about the flexibility of the planned charging session, such as planned arrival and departure times, required amount of energy and whether they allow vehicle-to-grid (V2G) . *Charge Point Operation* supports Charge Point Operators (CPO) to operate their CPs to fulfil charging requests relayed through EMPs. It needs extensions to support advance booking of CPs. *Roaming* enables seamless access to CPs by connecting EMPs and CPOs. It needs to be extended with support for advance booking and

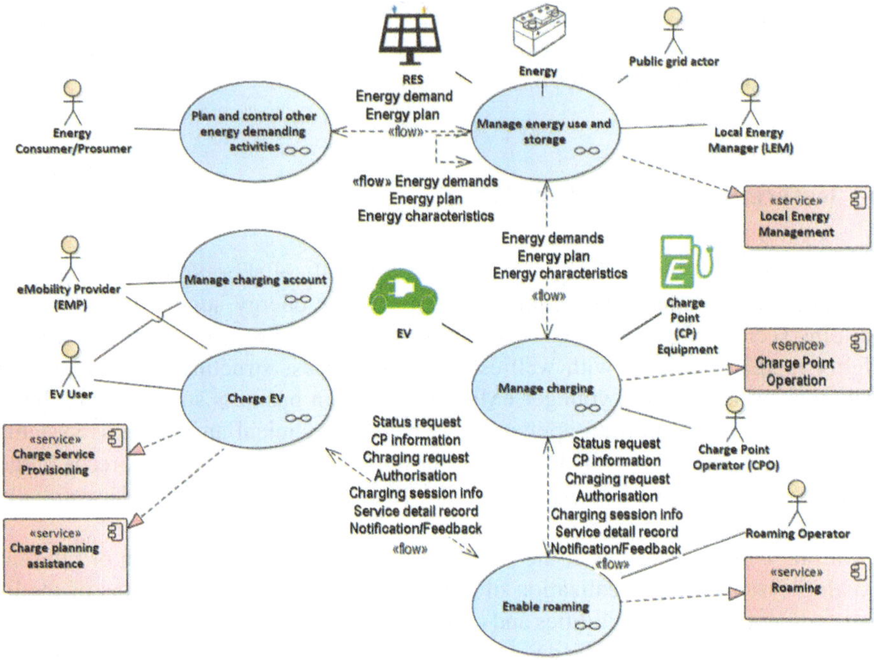

Fig. 2 Stakeholders, use cases and services for smart and green charging

for relaying the flexibility information. *Local Energy Management* offered by Local Energy Managers (LEM) automates shifting and shaping of flexible loads and use of local storage resources in a neighborhood with the aim to optimize utilization of local RES and reduction of peak power. Demand flexibility information is also solicited from the building inhabitants (energy consumers). Establishing such ESNs is a shared responsibility of the building owners in the neighborhood.

In GreenCharge, each ULL implements a selected subset of services defined in the RA according to the local context and business needs. Altogether they cover all the measures described in the RA.

5 Planning and Implementation of the eMobility Demonstrators

5.1 Planning of the Oslo Demonstrators

The real-life demonstration of eMobility in Oslo has been performed in a housing cooperative, Røverkollen, located in a suburb in the eastern part of Oslo, which

Table 1 Status of the infrastructure for the demonstration before and at the end of the project

Infrastructure	Before the project	At the end of the project
CPs	No CPs inside the garage 4 CPs outside the garage	64 CPs have been installed inside (46 CPs activated, and 30 CPs actively included in the evaluation)
PV	No installation	PV panels with 70 kWp capacity installed on the garage roof and work since October 2019
Stationary battery	No installation	A stationary battery with 50 kWh capacity installed on the ground floor, worked since the start of 2020, but stopped working since summer 2021
EVs	15	47

has 246 apartments spread on 5 housing blocks. Each apartment has its own, private parking space in a common four-floor garage. Table 1 shows the status of the infrastructure for the demonstration before and at the end of the project.

Before GreenCharge, 4 CPs outside the garage were used by the few EV owners in Røverkollen. A simple manual booking system based on an excel spreadsheet provided necessary coordination and predictability. With the increasing charging demand from new and potential EV owners, the housing cooperative decided to allow EV owners to install their own private CPs in the parking garage. This required additional cabling and strengthening of the local energy supply system. In Norway, a high peak load tariff is imposed for large consumers such as Røverkollen. Charging of many EVs simultaneously in the garage may cause high peak loads, leading to expensive energy bills and a high pressure to the existing electricity infrastructure. Røverkollen thus wants solutions for smart energy management which can facilitate the charging of all EVs and reduce energy peaks, and at the same time, ensure that the total charging demand is within the electricity capacity limits to avoid the expensive investment for upgrading the electricity infrastructure in the garage. Furthermore, the 4 outdoor chargers may also be shared by the public, such as visitors to the residents in the housing cooperative, utility vehicles visiting the area, employees and visitors at the nearby school, and any other EV user in the area. The sharing of these CPs may bring income to the housing cooperative.

The GreenCharge innovations offer solutions that can satisfy Røverkollen's needs. The original plan was to organize the whole housing cooperative as one ESN, which coordinates the energy management of the CPs inside and outside of the garage and the energy consumption in the buildings, including that of the apartments. However, the garage, the CPs outside the garage and the apartments in the buildings all have their individual energy meters and contracts with the retailers. To test the whole housing cooperative as one ESN in real life would thus require that all the energy entities in Røverkollen are under the same *virtual meter* to ensure fair distribution of the cost and benefits as well as the impact of the coordinated energy management. This is however not possible due to current regulations. Therefore, two separate demos have been set up to demonstrate the GreenCharge concept, each focusing on a selected set of measures: Demo 1 (Charging in ESN) demonstrates smart charging with energy management inside the garage, while Demo 2 (Advance booking of CPs) demonstrates roaming and sharing of CPs outside the garage. Business aspects have also been planned for each demo. Tables 2 and 3 show the planned measures and their expected output for Demo 1 and Demo 2, respectively.

Regarding the recruiting of participants of the two demos, residents were incentivized by the offer of discount of installation of a home charger at the garage with a load balancing add-on. The participants have a wide range of vehicles, with battery capacities from 8.8 to 82 kWh. Besides data automatically collected by software systems, the participants have completed a survey at the beginning of the project, and interviews at the beginning and at the end of the project, plus a focus group discussion with involved stakeholders.

5.2 *Implementation of Demo 1*

The Oslo Demo 1 demonstrates flexible charging with smart energy management and supporting business model in the garage as described in Table 2. The garage is managed by a software prototype implementing a selected subset of services defined in the RA adapted to the local context as illustrated in Fig. 3. The prototype is composed of four system components that realize four services defined in the RA. The *ZET App & Backend* realizes the *Charge Planning Assistance* and the *Charge Service Provisioning* services. EV users use the ZET App to start charging, provide input on user profile information (such as information about the EV) and to check the price for charging. In addition, users can select flexible or priority charging and specify the charging constraints in terms of expected departure time, initial SoC and target SoC. Default values can be saved in the user profile to simplify user input. The EV user can monitor the charging process and check the estimated SoC in the App. The *Local Energy Management* service is provided by the *eSmart Connected Prosumer* platform and the *ZET Individual Charge Planning*. The eSmart Connected Prosumer monitors and predicts issues that may affect the energy availability and use (weather, RES production, stationary battery, heating cables and charging demands), and calculates a dynamic *overall capacity plan* for cost-optimal energy use at an aggregated level (i.e., predicted total power capacity in the garage that can be used for the next 48 h with 15 min interval). The eSmart system also controls the charging/discharging of the stationary battery. When receiving the capacity plan, the ZET Individual Charge Planning takes the charging demands from the individual EV users into account and generates *individual charge plan* for each CP. The *Fortum Charging System* realizes the *Charge Point Operation* service for the control of the CPs. It controls the start/stop and the energy transferred at individual CPs according to the individual

Table 2 Demo 1 (Charging in ESN): status before the project, measures to be implemented and expected output

Category	Before the project	Measures to be implemented	Expected output
Energy management	• **No smart energy management**. All EVs start charging as soon as they plug in. The Load Management System (LMS) from the CP provider provides reactive control and allows the available energy shared equally among all plugged-in EVs	• **Smart energy management** supporting predictive optimal distribution of the energy use over time coordinated for charging and other energy use, adapted to energy availability, individual charging demands, and time of requested energy	• The distribution of available energy is fair and adapted to individual needs • A predictive cost-optimal energy management plan that reduces the power peaks of the garage and enables to charge more EVs at low cost • Smart use of energy from local RES/stationary battery storage makes the share of energy greener
Charging	• **No support for desired charging behavior**. Users have to manually postpone the charging to a time when the load or the price is low, by plugging in the EV later, or defining the charge plan setting in the EV itself (most EVs offer this possibility in settings). Not possible to change the charging current	• **Flexible charging** is default. EVs may be charged at any time before the user-defined latest finish time • **Priority charging** is optional. Priority sessions will be satisfied at the expense of non-priority ones in case available energy is not sufficient to satisfy all charging sessions	• New CPs in the garage makes charging easy and available for residents • Flexible charging arranges for smart energy management • Increased share of electric vehicles, leading to reduced CO2 emissions
Business aspects	• **No economic incentives for desired charging behavior**, e.g., to shift charging to periods with high production from local solar panels and/or low demand on the grid	• An economic penalty for priority charging: an extra fee charged (NOK/kWh) in addition to regular charging price for flexible charging (NOK/kWh) • Reward flexible charging for lower energy cost • Adjusted consumption pattern will be rewarded due to lower peaks or use of energy outside peak hours • Local RES is economically rewarded by means of a positive selling price if exported to the grid (equal to the market spot price), or by reducing electricity purchase by self-consumption	• The return of investment is expected to be covered by a profit from the payments from the EV users. The profit equals the difference between the income from the fixed charging fee (NOK/kWh) and the actual cost of electricity (paid to the electricity provider and grid owner) which depends on the market spot price and the garage's peak load • Reduced operational energy cost is expected through the utilization of local RES, reduced grid peak cost and reduced electricity cost through adjusted charging patterns

Table 3 Demo 2 (Advance booking of CPs): status before the project, measures to be implemented and expected output

Category	Before the project	Measures to be implemented	Expected output
Charging	• The residents had no private CPs in the garage, and had to use the 4 shared outdoor chargers and book the CPs via a spreadsheet • The shared CPs were only available to residents. Other EV users had no access to them. No mechanism supported the sharing of the CPs with the public	• The old CPs are replaced by new CPs and software, which arrange for data collection and billing • An App is used for CP booking, start of charging and payment. The booking may take place a short or a long time in advance (days) • Everyone can download the App and book and use the shared CPs	• 4 shared CPs are installed and available to the public • EV users can book charging time slots in advance and get predictable access to charging. The charging anxiety can be reduced • Drop-in charging is available in periods when the CP is not booked by booking a time slot starting immediately
Business aspects	• The residents paid a fixed monthly fee to the housing cooperative for use of the four shared CPs (located outside the garage)	• Payment for shared CPs: CP owner is compensated for offering CPs to others • Penalizing blocking of CP: EV users not using booked time slots (no-show or late arrival) or connected after booked time slot ended (blocking) are penalized	• The housing cooperative receives payment for the actual use of the CPs, which is used for return of investment • The price models encourage a desired behavior (e.g., in-time cancellations and avoiding blocking CPs) and compensate the housing cooperative in case of no-shows and blockings

charge plans. In addition, it is responsible for the EV user billing and payment which is part of the responsibility of the EMP role.

Since the Fortum Charging System and the eSmart Connected Prosumer system are commercial systems, special design and workaround has been done for the integration of the commercial systems into this prototype in order to reduce the changes of the operational systems, leading to the new development of the ZET components. As there is no open interface to get SoC information from the EVs, the ZET App is used to manually collect the SoC information needed for the eSmart system to predict the charging flexibility. Charging can however be initiated by an RFID chip if the EV user does not use the App for any reason. This is a workaround designed for user convenience (not to disrupt the charging if something goes wrong) and to lower the threshold for the trial as the software development and extension is done in

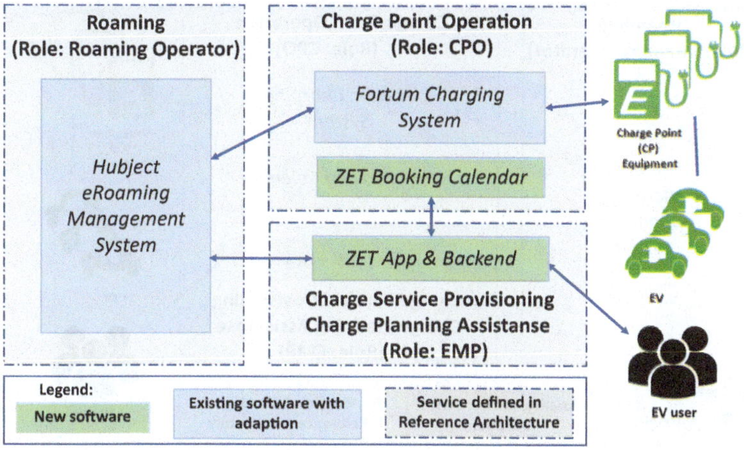

Fig. 3 The system components for Demo 1 and their interactions. The boxes with dash lines indicate the implemented services and their corresponding roles as defined in the RA

parallel with the normal use. In such a case, the EV will be charged with a minimum of current (8A) for 6 h. This is to motivate the use of App (e.g., for data collection and better smart management) while improving user acceptance.

5.3 Implementation of Demo 2

The Oslo Demo 2 demonstrates sharing of CPs through calendar-based advance booking and roaming with supporting business aspect measures as shown in Table 3. There is no business contract between the CPO (Fortum) and the EMP (ZET), but both have a contract with the Roaming Operator (Hubject). Visitors or people working in the neighborhood can book and use the four outdoor chargers owned by the housing cooperative using the ZET App without the need of an existing subscription with the CPO (roaming). The EV user can use the ZET App to find an available time slot in the future at a CP and to book a charge session. The booking defines the expected time slot and the energy request (defined in terms of initial SoC and target SoC). Bookings can be cancelled before the charging starts. Upon arrival at the CP, the user uses the App for authentication and plugs in the EV. The roaming platform authorizes the charging and supports the exchange of charging data, including the exact session time and energy charged.

Business and price models are designed to allow the CP owner a proper income with sharing CPs. Using price per hour instead of per kWh charged ensures a fair income, while extra fees for blocking the CP and no-show are defined to avoid blocking, booking deviations and inconvenience to the CP owner. To support the business aspects, the EV user is informed about cancellation and no-show conditions

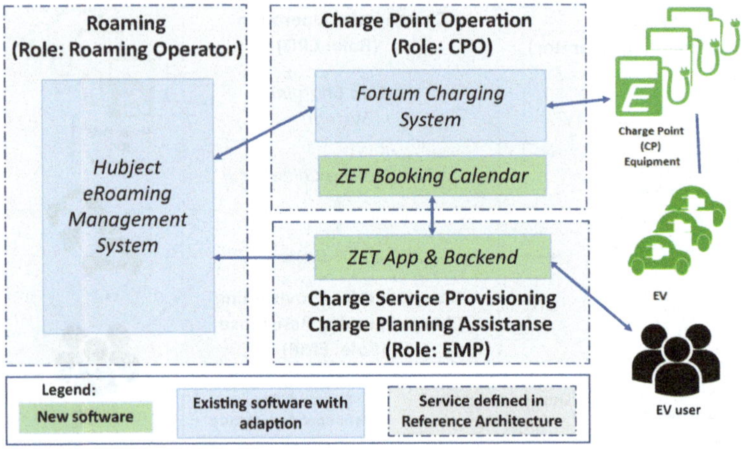

Fig. 4 The system components for Demo 2 and their interactions. The boxes with dash lines indicate the implemented services and their corresponding roles as defined in the RA

as well as blocking fee via the App. A payment reservation on credit card is done when the user makes the booking to cover the cost in case of no-show. If cancellation is done too late (such as less than 1 h before the booked time slot), 1 h fee must be paid. In addition, actions are taken to enforce the desired behavior and these measures. For example, the charging will be stopped if the EV is not unplugged by the end of the booked time slot. Notifications are sent to the EV user 15 min before the start of the reserved time slot (to remind on the potential no-show payment and cancellation deadline), 15 min before the end of the time slot (to remind on the end and the blocking fee), and if the EV blocks the CP after the end of the booked time slot (to remind on the blocking fee). Such notifications may reduce the number of no-shows, delays and blockings, and thus increase the acceptance for the CP owner and EV users.

Figure 4 illustrates the system components of the prototype implementing the measures defined for Demo 2. The prototype consists of four components realizing four services defined in the RA. The *ZET App & Backend* implements *Charge Planning Assistance* and *Charge Service Provisioning* services. It is basically the same component as used in Demo 1, but extended to realize the notifications described above and to work together with the new component *ZET Booking Calendar* to support advance booking as well as billing and payment functionality. The *Hubject eRoaming Management System* offers *Roaming* services, which supports authentication and authorization as well as connects the CPO system provided by *Fortum Charging System* and the EMP system provided by ZET App & Backend.

The Fortum Charging System and the Hubject eRoaming management System are commercial systems. Similar to Demo 1, special design has been done for the integration to reduce the changes of the operational systems, leading to the new development of the ZET components. In particular, according to the RA, ideally the booking shall be supported through the roaming protocol so that EMPs can get an overview of a calendar-based CP availability from CPOs. The current commercial roaming protocol (OICP protocol 2.2) only supports the "reserve now" functionality, and does not support advance booking. Our solution is to develop the ZET Booking Calendar component as an extension of the CPO functionality to implement the calendar system supporting the advance booking of charging sessions. In addition, the OICP protocol for roaming has been extended to get information related to the charging and billing, e.g., about the exact session time and energy charged. In such a design, the ZET components have implemented both the EMP and part of the CPO functionality. In the future, a standardized interface for advance booking and for relaying the energy demand and flexibility information with a clearer role definition of EMP and CPO shall be provided by an extension of the roaming protocol in order to ease the integration and interaction between EMPs and CPOs.

6 Evaluation and Lessons Learned

All planned measures for Demo 1 and Demo 2 as described in Tables 2 and 3 were implemented, but some of them were not operational due to COVID-19 and delays in the deployment. In spite of this, the impact on stakeholder acceptance and awareness and lessons learned from the innovation process can inspire others who want to implement similar services.

6.1 Impact on Stakeholder Awareness and Acceptance

This section presents the impact evaluation findings regarding the stakeholder awareness and acceptance of the measures implemented in Demo 1 and Demo 2.

6.1.1 Demo 1: Charging in ESN

The situation before the project described in Table 2 is compared with the situation when the measures described were operative. The evaluation is based on input from a group interview with the housing cooperative board, interviews with selected residents, surveys, and an interview with the chairman of the board after the testing of the App. Generally, high awareness and acceptance of the new measures is indicated. **Charging**. Awareness among residents and EV users improved from low to high during the project as a result of several information meetings and abundant dissemination material. In general, the acceptance of eMobility has been high both before and during the project. Regarding flexible charging and the use of the new App for charging, early in the demo development phase, there were questions about "what's the point of the App?" and "why it is so expensive to charge?". No further questions were asked after the residents were informed about the positive effects of the App on the energy use (how the use of the App may influence the peaks in energy use and costs). The high energy price late in the project[6] has also significantly affected the high acceptance of the App. Furthermore, compared to the previous solution, the new system is easy-to-use and perceived as more convenient, thus the charging infrastructure is considered as making EVs more available and increasing the accessibility for the residents.

Smart energy management. Residents are not so aware of the existence of PV panels as they are of the new charging infrastructure. However, it is stated that the smart energy management is seen as a positive service. It can among others increase the knowledge and environmental awareness among the residents and may have a positive impact beyond just the residents with EV. The contribution to lower electricity bill is also welcomed. The annual report communicated that yearly PV production was 55000 kWh. PV production lowers the bill and unused energy is sold back to the grid, leading to an income to the housing cooperative.

Business aspects. The housing cooperative was involved in the elaboration of the business and price models. Thus, the awareness and acceptance is high. Many iterations are needed to come up with an understandable and acceptable proposal for the business and price models. The prices for charging are in general low as the aim of the housing cooperative is to cover the operational costs and minimize the costs for the residents, rather than earning money. Further, economic incentives are a driver for acceptance of flexibility and a mechanism for discouraging priority charging. The main challenge is to decide how to reward the desired charging behavior. According to the interviewees, the compensations for flexibility need to be significant in order to cause the desired effect.

6.1.2 Demo 2: Advance Booking of CPs

The measures described in Table 3 have been implemented, and it is tested that they work. Delays have however prohibited the roll-out of the demo. Despite of not being operational, the implementation process has facilitated important learning. Table 3 presents the situation before the project and the expected output of the new measures. Generally, there is a high awareness and acceptance of the measures among residents and the housing cooperative.

Charging. Residents are aware of the existence of the outdoor charge points and are willing to recommend them to visitors. The housing cooperative wants to offer charging to visitors and EV users in the neighborhood. The service is thus established in close collaboration with the housing cooperative and gets high acceptance from them. The expected acceptance among external EV users could not be measured, but is assumed to be high as there are few public charge points in the area, advance booking could facilitate predictable access to charging, and prices are low.

---

[6] The energy prices in Norway have increased with several hundred percent in 2021, and it is expected to remain high.

Business aspects. The housing cooperative was involved when elaborating the business and price models. Thus, the awareness and acceptance is high. It is emphasized that the price model should ensure a revenue also in case of undesired charging behavior such as no-show and blocking. The penalties for such behavior are thus set to be high enough to ensure that. The EV users' expected acceptance is assumed to be high since the price level is lower than at other public chargers and the users will get predictable access to charging in addition.

6.2 Process Evaluation Findings

This section presents the findings from the process evaluation of the two demos.

6.2.1 Key Drivers, Barriers and Supporting Activities

Table 4 provides an overview of the key drivers identified and their effects. Table 5 provides an overview of the key barriers identified and actions to overcome them. Detailed description of the complete list of drivers and barriers can be found in [15].

The biggest barriers we experienced are associated with the lack of standardized protocols for the novel measures of GreenCharge solutions and the difficulties with making experimental modifications and extensions to 24/7 systems. The integration with CP equipment became a challenge. The built-in CP Load Management System (LMS) with simple reactive load balancing blocked the scheduling done by the smart energy management. At the end, the built-in LMS had to be disabled. However, this led to a safety risk in case of technical problems (e.g., software errors, loss of Internet connection) where fuses may blow, equipment may be damaged, and charging may be blocked. To mitigate this, a software workaround was designed regarding smart energy management: a low speed default charging (8A) ensures some charging in any case, and capacity is reserved for low-speed charging, to prohibit overloading. The disadvantage of this approach is that the reservation leaves less capacity for use in optimizations, and the value of the solution is reduced.

Another serious barrier is that the electric energy sector is a highly regulated one with a regulatory regime lagging behind the rapid growth of decentralised RES and the increasing interest in local energy communities like ESNs. In Røverkollen, although it was technically feasible to create a virtual meter and bill the initially planned ESN as one entity, it was refused by the local Distributed System Operator (DSO) because it was against the regulation. We also suspect that DSOs are reluctant to accept such solutions because they would lose payment for the use of the grid to transfer locally produced energy between the members of the local community. Hopefully, this will change in the future as local RES and smart house technology become more widespread.

In addition, the following key supporting activities have contributed positively to the process:

Table 4 Key drivers observed for Demo 1 and 2 and their effects

Driver types	Drivers	Effects
Behavioral	**Positive attitude** from the housing cooperative administration to the new solutions	Input on needs and possibilities. Leaders involved in decisions on App functionality, business models and risks
	Subsidies of CP investments	Participants recruited
	Promotion activities such as launch event, explanatory videos, information meetings and letters to residents	Increased awareness and knowledge among residents
	Stakeholder engagement through meetings and workshops on business models	Roles, focus and user needs clarified (income, penalizing priority, alignment with price levels in general, etc)
	High eMobility acceptance in Norway	Easy to recruit
Economic	**eMobility incentives** in Norway, such as tax reductions, toll road fee reductions	eMobility acceptance and positiveness
	Subsidises from municipality and housing cooperative association	Investments in CPs, stationary battery, and PV panels. The demonstrator of this size would have been impossible without the subsidises
Impl. capacity	**Multidisciplinary team** with technology providers (CPO, EMP, LEM), business model experts, municipality, researchers, housing cooperative administration, and residents	Cross-sectoral solutions, in line with the ULL characteristics
	Flexibility of partner with respect to additional tasks and responsibilities	Problems could be solved
	Weekly follow up meetings for demos	All involved partners participated, blockers identified, tasks assigned and followed up
Technical/economic	**Business models and technology alignments** in workshops with business model designers and technology providers	Technology support for the implementation of business models that motivate desired behavior (e.g., booking enforcement, extra fee on priority charging)

Table 5 Key barriers observed for Demo 1 and 2 and actions taken to overcome them

Barrier types	Barriers	Actions to overcome
Complexity	**Complex and novel solution**. Lack of off-the-shelf components and integration of systems from several partners. Time and resource demanding testing, debugging and bug-fixing	Requirement updated to match new insight. Unforeseen problems addressed when they occurred
Technical	**Integration with CP equipment.** Interface did not work as expected and blocked the smart energy management	Testing, investigations, and delays. Software workaround solved problem and safety risk, but reduced the quality
	Integration of outdoor heating cables. Difficult to control correctly. Cannot predict the consumption pattern due to unknown dependencies on temperature, delay, humidity, etc	Direct control not implemented due to safety (cannot risk icing) and technical problems (high power requires expensive equipment). Integrated as a non-flexible residual load instead
	Integration between management systems. No standardized interfaces for integration of charge management and LEM	Custom interfaces to the systems and their capabilities (as planned)
	Access to SoC is a problem since current standardized protocols do not support access to current SoC from the EV's on-board systems	Software workaround: App asked EV User to provide data on the initial SoC as well as the desired SoC. The initial value might however be inaccurate
	Booking principles. Scarcely previous work, no existing systems with intended solution. No existing standards	New functionality and new coordination mechanisms designed and implemented
	Roaming. Combining booking and roaming was more complex and challenging than expected. Standards lack booking support	Focus on the implementation of the booking functionality to facilitate learning and to specify novel solutions in the RA
Impl. capacity	**Limited testing ability**. The outdoor chargers have been the backup for the garage in Demo 1 during construction work in the garage for several months towards the end of the project and delayed the roll-out of the demo	Postponed activity till after the construction work. Demo 2 did not start operational during the project period
	Administrative issues. Onboarding into the roaming platform, the opening of APIs, and contractual issues were not accounted for and took much more time than expected and became blockers	Weekly telcos to follow up the status and blockers. Bug-fixing in contracts. Involvement of leaders. Late roll-out of the App.
	Operational system challenges. The new measures could not be easily integrated in systems that already are in operation	Software workaround. New functionality was implemented in new components

(continued)

Table 5 (continued)

Barrier types	Barriers	Actions to overcome
Regulatory	**Regulatory barriers of current grid regulations.** Demands each main meter to be billed individually. The DSO did not consent to the creation of a virtual smart meter for all the buildings of the housing cooperative as a whole, thereby blocking the billing of the planned ESN as one customer. Hence, the smart charging was used to reduce the peak load of the garage solely, instead of the whole housing cooperative	Data from the garage and from some apartments were used as input to simulation of more comprehensive ESNs in order to investigate the possible effects on key performance indicators
Collaboration	**Coordination and communication problems** due to the multidisciplinary work and the many actors involved as well as COVID-19 impact (sickness, low priority internal to commercial partners, etc.)	Weekly telcos to follow up blockers and coordination between activities. Experts within various areas (energy, hardware, software, integration and development process) worked as a team and explained problems in an understandable way
Economic	**Traditional business models** addressed the commercial actors and their income and did not represent innovation (too generic, not using the opportunities created by the technology, no incentive to the desired behavior)	Re-design of business models focusing on the housing cooperative and the economic effects of the ESN

Involvement of the housing cooperative administration and residents. The housing cooperative administration and residents were involved through workshops, meetings and communication initiatives like videos, information letters, and launch events. Input on needs, possibilities and views upon flexible charging, priority charging and other aspects as well as input to the acceptance evaluation were collected through interviews and questionnaires. In the last part of the project, there were frequent meetings with the housing cooperative administration on a weekly or daily basis to exchange status information, plan testing activities and solve problems. The good relationship with the housing cooperative and their positive attitude to the new solutions contributed positively to the planning and implementation of the demos.

Daily (core group) and weekly (larger group) follow up meetings. All partners involved in the implementation and roll-out of the demonstrator participated. The meeting contributed to a tight follow up of the demos. Problems were identified and addressed at a detailed level. Decisions on actions could be taken more quickly since all partners with a role were present. Moreover, the expert participants communicated good about complex problems and created a common understanding that inspired dedicated work to solve problems.

Workshop on business models. The main actors in the business models and the partners for the technical development participated to re-design the business models and to decide the price models. The approach was a success as it clarified the money flows between actors and facilitated concrete discussions regarding decisions on prices (for example, prices for comparable services, economic incentives for desired behavior, how to ensure income in case of blocking) and the actual prices.

Live testing at the demo location. The project had arranged 5–7 test events for testing of EV charging and related software and hardware. As participants from all partners involved in the implementation participated and experienced and identified errors through real-time debugging, responsibilities and actions to fix the problems could be easily agreed.

6.2.2 Key Lessons Learned

Key lessons learned important for future sustainable eMobility strategies include:

Flexible charging can be implemented, provided that the charge management system and the CP equipment can be integrated and controlled in a detailed and flexible way. CP equipment control must support control of individual CPs, i.e., to start and stop the individual CPs and to charge with different power as desired.

The ESN implementation is a big challenge. This can not be done by easy plug-and-play currently since off-the-shelf components from various providers (among others CPO and LEM) cannot easily be integrated due to the lack of standards and standardized interfaces. Furthermore, it may also be difficult to control the systems and involved equipment, CPs included.

Demonstration of innovative measures is more complex than foreseen. In particular, the administrative issues linked to the opening of APIs, contract, and the onboarding into the roaming platform took much more calendar time than expected and blocked progress for Demo 2 for a long time.

Business models should address more than just the money flows. One mechanism is to use price models to encourage a desired behavior. For example, flexibility should be rewarded (this is not common today) and blockings, no-show or cancellation must be penalized.

7 Discussion

7.1 Recommendations

The following recommendations can be derived from the process evaluation.

Stakeholder involvement. Several types of actions must be considered to get input from and to involve the stakeholders to increase the stakeholder awareness and acceptance, for example, workshops and meetings, information letters to EV users, launch events, interviews and questionnaires. Examples to increase the awareness of the shared CPs could be information letters to contact points for EV users that may use the CPs (e.g., other housing cooperatives in the area, and a school in the area) and information provided to the eMobility association so that they can promote the CPs via their channels (website showing CPs, newsletters, etc.). Affected stakeholders must be involved whenever relevant, for instance, regarding the purchase of hardware and the design of the functionality supported by the technology (such as the App functionality) as well as the business models and price models. It is also important to ensure that users know how they can find information and get support.

Design and implementation. The coordinator/leader of the demo activities must have high technical expertise and multidisciplinary experts need to be involved. It is not always easy to detect potential problems regarding software and integration. For instance, the energy management issues must be understood and the software knowledge is needed to follow up and interpret the status reporting from the technical partners. In addition, the implementation must be followed up regularly (e.g., at a weekly basis) and all partners involved must participate. Blockers and problems must be identified at a detailed level. Actions must be decided, responsibilities must be assigned, and blockers and actions must be followed up. Administrative and formal issues (such as roaming onboarding, opening of APIs and contracts) must be planned and addressed so that they do not delay the implementation. The needed administrative activities with their estimated time for carrying out must be identified early, and planned and handled in parallel with the technical implementation.

Hardware and equipment for ESNs. The needs must be specified in detail. The lessons learned show that the ability of the devices stated by the providers cannot be fully trusted unless they are based on a detailed specification of the needs or come from those with detailed technical knowledge. In particular, the integration with the energy management system and the ability for equipment control (for instance, the ability of software to control the CPs) must be emphasized and verified. Today, CPs are often provided with a built-in solution for simple load balancing, which does not allow the local energy management in the ESN to control the CPs as required and may cause problems in an ESN. Thus, the details must be discussed with the provider of the software controlling the CPs to facilitate integration with the energy management system. Considering the problems described above, the involvement of experts is crucial and most property owners should use external expertise for the design and development of the total ESN solution.

Business models and price models. The business and price models must be designed in collaboration with all the involved stakeholders. The property owners (like the housing cooperative) must be involved in the decisions and their needs be addressed. The traditional approach to business models is not adequate. The value proposition is also about charging behavior and sustainability with respect to environmental and societal aspects, such as willingness to offer flexibility and reduction of energy peaks. Suitable business models and price models with supporting technical solutions have the potential to motivate to a desired behavior (such as avoidance of no-show or blocking of CPs) and to handle business related problems.

Policy, standardization, and harmonization issues. The software integration between local energy management and charge management must be standardized. Charging protocols must provide the current SoC to facilitate optimal charge planning in ESNs and EVs should provide open APIs for the access of the SoC information. For instance, manually entering SoC data via App is not an ideal approach due to inaccuracy and extra effort for EV users. Automated SoC reading is needed, but it demands that EV producers open APIs for access of such data based on standards. Navigation systems must support automated charge planning and booking that can adapt the charging constraints to the needs, for instance, based on planned trips or artificial intelligence using input on the EV user's habits. Providers of CP equipment must facilitate the integration with local energy management in ESNs to support an extended load balancing based on predictions and the needs of the whole ESNs. Providers of devices like stationary batteries must recognize the needs in ESNs and support the required control mechanisms.

7.2 Deployment of Smart Charging in Large Urban Areas

One big barrier to the large scale deployment of smart charging is the lack of standardization as discussed in Sects. 6.2.1 and 7.1. Without standardization and harmonization of interfaces and protocols for control of CPs, interaction between different system components and access to accurate SoC information, solutions will not work simply by plug-and-play as the components will not be easily integrated, and it will be difficult to predict the charging needs accurately.

One key advantage of smart charging with smart energy management as implemented in GreenCharge lies in the *optimization of the total energy cost*. The advanced local energy management from GreenCharge seeks to charge the EVs at the lowest possible cost and with the greenest possible electricity and takes into account the peak load tariff, variable electricity sport prices in the electricity market and the locally produced solar energy. The solution will move the energy use (charging included) to times when the varying price costs are low and at the same time lower the peaks (thus reduce peak power payment) and maximize the utilization of the locally produced solar energy (free of use and at the same time reduce the electricity bought from the power grid). The cost saving potential is thus big, even for periods when the PV production is low and the total energy use is high.

Although there is no comprehensive evaluation of the economic impact of the demonstrated solutions due to the short operative period of the new implemented measures, a few observations are noticeable for the demos. Through the revenue from export of surplus solar energy, the PV panels have reduced the cost for the provision of the charging services in Demo 1 by 4512 NOK with 0.151 NOK per kWh for a period of 6 months. This corresponds to a reduction of about 10% of the average charging cost. However, such profit varies over the year, in particular, the PV production is low in November and December when there is little sunlight and the PV panels are covered by snow.

Even though the GreenCharge solutions do not explicitly try to optimize the utilization of the charging infrastructure, our proposed advance booking mechanism will tend to spread the charging traffic to charging stations with free capacity, which contributes to better utilization of charging infrastructure and reduced queues. Moreover, sharing of private CPs via advance booking will bring economical benefits to CP owners, which can motivate the sharing of CPs and increase the availability of charging infrastructure. Thus, the need for overprovisioning of charging capacity in large urban areas will be reduced.

The investment on the charging infrastructure, the stationary battery and the PV installation in the demos has been well subsidized and contributed very positively to the deployment and acceptance of the solutions. Such investment represents a fair amount of cost for starting-up and should be considered in large scale deployment. In addition, investment cost related to system integration will be substantially higher without standardization. With smart charging, the return of investment for PV panels will improve especially with the high energy prices. Further, governmental support or subsidy can facilitate wide scale deployment.

For both cases of booking (*reserve now* and *advance booking* as introduced in Sect. 1), it is difficult to fully guarantee the availability of a reserved CP to the user who reserved it. Although the CP owner may limit the use of the CP in the reserved period to the user who reserved it, it is still possible that somebody blocks the associated parking space. This is more likely to happen with advance booking, since the reserved CP may be used by others in the period between booking and start of the reserved period. Possible remedies are to fine misbehaving users and/or use remotely controlled physical barriers. GreenCharge adopts economic penalty in the business model and App notifications to discourage users with bookings from blocking of unused CPs, but other measures are needed to prevent the blocking of CPs without bookings, such as remotely controlled physical barriers (which may be too expensive though).

7.3 Validity of the Results

The eMobility acceptance in Norway is very high, and this has probably impacted the stakeholder acceptance evaluation. A new legislation that dictates that all housing cooperatives must facilitate infrastructure for EV charging may have influenced the acceptance in a positive way.

The subsidies that residents and the housing cooperative received for the CPs and PV panels may also have affected the acceptance and the awareness. The fact that all charge points are not activated also indicates this, but at the same time confirms a high degree of eMobility awareness and acceptance. Many residents have concrete plans for a replacement of their fossil vehicles with electric vehicles.

Demo 2 measures have been implemented and tested, but not operational, so the stakeholder acceptance evaluation of Demo 2 should be considered as expected awareness and acceptance from potential users. Advance booking of CPs is a new service that mainly is relevant when the eMobility acceptance is high and the number of EVs is high (and thus the pressure on charging infrastructures is high). Therefore, the findings with respect to the expected acceptance and awareness are relevant even though the shared outdoor CPs have not been operational.

Most findings and lessons learned are common for both demos and may also be common for other eMobility use cases. Since Norway is a world leader in adoption of EVs, the evolution and deployment of eMobility in the Oslo demos may be representative for future evolution of eMobility and provide learnings for other countries.

8 Conclusion

The GreenCharge Reference Architecture provides a full-fledged specification for implementing solutions for smart and green charging. It has identified the needed extensions of existing charging protocols and stakeholder roles to support the novel solutions like advance booking of CPs and integration with smart energy management in a neighborhood. When implementing and demonstrating such solutions in selected urban living labs, adaptions and workarounds have to be done due to the constraints of the local context and the available technology. Moreover, a successful adoption of innovative solutions needs not only well-functioning technical systems, but also support in other aspects, such as user engagement and economic and policy incentives.

There are few reports on realizations and experience of adopting innovative and sustainable eMobility solutions in real life. This chapter attempts to bridge the gap and presents the implementation and demonstration of smart charging integrated with smart energy management and advance booking of CPs for sharing private CPs in a housing cooperative in Oslo as well as lessons learned from this process. We experienced barriers related to technical, regulatory and collaboration issues, in particular, the integration with the energy management system, the ability for equipment control, the lack of support in commonly used protocols for the proposed innovative measures, and the difficulties with making experimental modifications and extensions to 24/7 systems. Recommendations to overcome such barriers are suggested. For instance, standards are crucial for facilitating the integration of local energy management and charge management, and open interfaces to access vehicle SoC are needed for automated and accurate prediction of energy demand and flexibility.

Further, stakeholder acceptance evaluations indicate a high acceptance and awareness of the solutions among the trial users in general. Our experience showed that good communication with users of the solutions and their expected benefits can improve user acceptance of flexible charging and motivate them to use the App and provide flexibility data needed for optimization. Due to the lasting increase in the electricity tariffs, the smart charging solution with the potential to reduce energy bill has got high acceptance by the housing cooperative, confirmed by their willingness to maintain the services after the project end.

Since advance booking of CPs is not commercially available yet, this chapter provides a detailed description of the expected functionality of a user App and the supported business and price models to encourage desired behavior. The solution may give inspiration to technology providers and service providers for the wider takeup of sharing of private CPs.

Demonstration with innovative eMobility solutions in real life was more complex than foreseen. Due to the short operational period of the Oslo Demo 1 and the delay of the operation of Demo 2, we were not able to fully evaluate the stakeholder acceptance of all the innovative measures and their full potential. Future work is needed to gather more empirical evidence on the impact of such measures through long period, large-scale trials.

Acknowledgements Authors of this chapter, on behalf of GreenCharge consortium, acknowledge the European Union and the Horizon 2020 Research and Innovation Framework Programme for funding the project (grant agreement n° 769016). The authors would like to thank all project partners for the collaboration and project management, in particular, ZET, eSmart, Fortum and Hubject for the implementation of the prototypes for Oslo ULL, EGEN/PNO for the business model design, and Oslo municipality for the coordination, as well as Røverkollen for the support and involvement.

References

1. Alexeenko P, Bitar E (2021) Achieving reliable coordination of residential plug-in electric vehicle charging: a pilot study. arXiv:2112.04559
2. Basmadjian R, Kirpes B, Mrkos J, Cuchý M (2020) A reference architecture for interoperable reservation systems in electric vehicle charging. Smart Cities 3(4):1405–1427
3. Dudek E (2021) The flexibility of domestic electric vehicle charging: the electric nation project. IEEE Power Energy Mag 19(4):16–27
4. Engels D, Bergh GVD, Breemersch T (2017) Refined CIVITAS process and impact evaluation framework. CIVITAS SATELLITE Project Deliverable D2.3
5. ENTSO-E (2021) Position paper on electric vehicle integration into power grids
6. Flocea R, Hîncu A, Robu A, Senocico S, Traciu A, Remus BM, Răboacă MS, Filote C (2022) Electric vehicle smart charging reservation algorithm. Sensors 22(8):2834
7. Hardinghaus M, Anderson JE, Nobis C, Stark K, Vladova G (2022) Booking public charging: User preferences and behavior towards public charging infrastructure with a reservation option. Electronics 11(16):2476
8. Hevner AR, March ST, Park J, Ram S (2004) Design science in information systems research. MIS Q 28(1):75–105
9. Iversen EB, Morales JM, Madsen H (2014) Optimal charging of an electric vehicle using a Markov decision process. Appl Energy 123:1–12
10. Jiang S, Floch J, Natvig M, Tricker R (2022) Summary of project achievements. GreenCharge Project Deliverable D1.4
11. Jiang S, Natvig M, Hallsteinsen S, Lindberg KB (2022) Lessons learned from demonstrating smart and green charging in an urban living lab. In: Barolli L, Hussain F, Enokido T (eds) AINA. Springer International Publishing, Cham, pp 624–636
12. Natvig M, Jiang S, Hallsteinsen S (2020) Stakeholder motivation analysis for smart and green charging for electric mobility. In: Barolli L, Amato F, Moscato F, Enokido T, Takizawa M (eds) Web, artificial intelligence and network applications. Springer International Publishing, Cham, pp 1394–1407
13. Natvig MK, Jiang S, Hallsteinsen S, Subramanian A (2022) Final architecture design and interoperability specification. GreenCharge Project Deliverable D4.2

14. Natvig MK, Jiang S, Hallsteinsen S, Venticinque S, Enrich Sard R (2021) Evaluation approach for smart charging ecosystem-with focus on automated data collection and indicator calculations. In: Barolli L, Woungang I, Enokido T (eds) AINA. Springer International Publishing, Cham, pp 653–666
15. Natvig MK, Sard RE et al (2022) Final result for innovation effects evaluation/stakeholder acceptance evaluation and recommendations. GreenCharge Project Deliverable D5.5 and D6.4
16. Orcioni S, Conti M (2020) EV smart charging with advance reservation extension to the OCPP standard. Energies 13(12):3263
17. Ottesen S (2021) The need for data and algorithms. Solenergi Webinar, 22 Jun 2021
18. Pantoš M (2011) Stochastic optimal charging of electric-drive vehicles with renewable energy. Energy 36(11):6567–6576
19. Quirós-Tortós J, Ochoa LN, Butler T (2018) How electric vehicles and the grid work together: lessons learned from one of the largest electric vehicle trials in the world. IEEE Power Energy Mag 16:64–76
20. Salpakari J, Rasku T, Lindgren J, Lund PD (2017) Flexibility of electric vehicles and space heating in net zero energy houses: an optimal control model with thermal dynamics and battery degradation. Appl Energy 190:800–812
21. Smart JG, Salisbury SD (2015) Lessons learned about plug-in electric vehicle charging infrastructure from the EV project and chargepoint America. Technical Report, Idaho National Lab (INL)
22. Steen K, van Bueren E (2017) The defining characteristics of urban living labs. Technol Innov Manag Rev 7:21–33

Shanshan Jiang received her Dr. Ing. (Ph.D.) degree in telematics from Norwegian University of Science and Technology (NTNU) in 2008. She is a Senior Research Scientist in SINTEF Digital, Trondheim, Norway. She has extensive experience with EU and national research and innovation projects within the area of software engineering, smart cities and smart energy management. She has been leading several EU and national research and innovation projects within the open data and smart cities areas. She was the technical manager of EU projects GreenCharge and CoSSMic (both focusing on energy smart neighborhoods). She has published more than 30 peer-reviewed papers in national and international conferences and journals. Dr. Jiang's research interests include software engineering and architecture, blockchain, open data and data management, ICT for smart cities and digital technologies for the green transition. She was the co-chair of the Second Blockchain Software Engineering Workshop (BSEW/EASE 2021). She has served as a program committee member for a number of IEEE/ACM international conferences since 2015.

Marit Natvig has a M.S. in computer science from the Norwegian Institute of Technology (NTH), in 1982. She is a Senior Research Scientist in SINTEF Digital, Trondheim, Norway. She was for 9 years the research leader of a research group on intelligent transport systems and smart cites, and she is also the winner of the national ITS price for work within intelligent transport systems. She has extensive experience from national research projects (also as the project leader) and from several European research projects. In the EU project GreenCharge she was the evaluation manager with responsibility for the evaluations methodology, the coordination of the evaluation activities, and the evaluation results, and she also was responsible for the architectural work. Currently, she is the technical manager in the EU project ORCHESTRA, dealing with future traffic management. Natvig's fields of expertise are software engineering, open data, innovation based on open data, smart cities, and digitalization within the transport domain. In total she has more than 90 publications of which more than 15 are peer-reviewed papers in international conferences and journals.

Svein Hallsteinsen has a M.S. in physics and applied mathemetics from the Norwegian Institute of Technology (NTH) from 1970. He is a senior scientist at SINTEF Digital, Trondheim, Norway. He has extensive experience from R&D projects in the domain of software engineering and has participated in a number of EU funded research projects where he has served both as technical manager and coordinator. Recent projects include CoSSMic and GreenCharge which both focused on energy smart neighborhoods. In GreenCharge he has contributed primarily on architecture and impact evaluation. Hallsteinsen's research interests include software architecture, software product lines, self-adaptive software, mobile and ubiquitous computing and the application of ICT to facilitate the green shift.

Karen Byskov Lindberg received her M.Sc degree (2005) and Ph.D. degree (2017) in Electric Power Engineering from the Norwegian University of Science and Technology (NTNU). She is an adjunct associate professor at NTNU, Trondheim, Norway, and a senior advisor in Aneo (aneo.com). Lindberg has extensive experience from the public sector (Norwegian energy regulator) and the research sector (SINTEF Community), where she has worked with long-term energy system analysis, energy policy, power markets, grid tariffs, energy performance of buildings, and development of local energy systems. Lindberg's research interests include energy efficiency of buildings, heat pumps, advanced statistics for load prediction, and optimal utilisation of local renewable energy sources combined with end-user flexibility in buildings and flexible charging of EVs.

Vehicle Allocation Algorithm Improving User Satisfaction in Ride-Sharing

Yuta Wakazono, Shigemi Ishida, and Yoh Shiraishi

Abstract Ride-sharing services have attracted attention as a new public transportation service. Previous studies on ride-sharing services focused on a vehicle allocation algorithm maximizing the number of accepted requests while minimizing vehicle travel distance. To make ride-sharing services sustainable, maximization of user satisfaction is important. Users have different satisfaction aspects. We therefore propose a vehicle allocation algorithm that can handle the individual user satisfaction preference. The key idea is to switch the cost function used in the optimization of vehicle allocation based on an individual user satisfaction preference. In this paper, we assume that there are users whose satisfaction preference consists of quick arrival and economy. We define convenience, economy, and balance cost functions based on user satisfaction preferences of quick arrival and economy. Combining the cost functions with the existing successive best insertion (SBI) vehicle allocation algorithm, we allocate vehicles to requests while maximizing user satisfaction. We show the improvement of user satisfaction compared to the existing vehicle allocation algorithm by up to 11% in simulation experiments a dataset of taxi trips on Manhattan Island in New York City.

Y. Wakazono
Graduate School of Systems Information Science, Future University Hakodate, Hokkaido, Japan
e-mail: wkznyt1126@gmail.com

S. Ishida · Y. Shiraishi (✉)
School of Systems Information Science, Future University Hakodate, Hokkaido, Japan
e-mail: siraisi@fun.ac.jp

S. Ishida
e-mail: ish@fun.ac.jp

R. Shinkuma et al. (eds.), *Advances in Engineering and Information Science Toward Smart City and Beyond*, Engineering Cyber-Physical Systems and Critical Infrastructures 5, https://doi.org/10.1007/978-3-031-29301-6_6

Acronyms

SBI	Successive Best Insertion
DARP	Dial-a-Ride Problem
SUMO	Simulation of Urban MObility
TraCI	Traffic Control Interface
SAVS	Smart Access Vehicle Service

Glossary

Ride-sharing service A ride-sharing system schedules the route of vehicles to allow users to share a ride providing low-cost door-to-door transportation.

User satisfaction A satisfaction with the service received by users of ride-sharing services.

User satisfaction preference A user satisfaction preference is represented by a combination of convenience weight W_C and economy weight W_E, where $W_C + W_E = 1$. The ratio of W_C and W_E is assumed to be unchanged from the pre-defined ratio.

Dial-a-Ride Problem The Dial-a-Ride Problem consists planning of vehicle routes and schedules for multiple users who specify the origin and destination points of the request.

1 Introduction

Recent advances in information systems and wireless communication technologies have led to practical ride-sharing services. In ride-sharing service, a ride-sharing system schedules the route of vehicles to allow users to share a ride providing low-cost door-to-door transportation. Ride-sharing services are one of the promising candidates for future transportation services [1].

To realize practical ride-sharing services, pioneering work studied the feasibility of ride-sharing services and optimization of vehicle routing as well as allocation [2, 3]. Noda et al. and Ma et al. showed the advantages of ride-sharing services compared to public vehicle-transportation systems such as buses and taxis.

Allocation optimization in dynamic situations where requests are sequentially generated is difficult because of a considerable amount of computation. In ride-sharing services, vehicles are allocated by a heuristic algorithm due to the difficulty in allocation optimization. The literature has reported the optimization of ride-sharing services based on estimated demands/requests, efficient vehicle routing as well as allocation, and various pricing strategies [4–9]. Sustainability is an important aspect to encourage users to continue to use a ride-sharing service as a public transportation service. However, the sustainability of ride-sharing services has not been well studied.

To realize sustainable ride-sharing services, maximization of user satisfaction is an important aspect. This paper therefore proposes a vehicle allocation algorithm considering the satisfaction of users. Our key idea is simple: we optimize vehicle allocation with the cost function considering user satisfaction preference. User satisfaction depends on an individual user's situation. We assume that a ride-sharing sharing system asks users to explicitly declare their requirements such as *quick arrival* and *the lowest cost*.

As a first step of the ride-sharing service considering user satisfaction, this paper presents a vehicle allocation algorithm maximizing a user satisfaction metric. The vehicle allocation algorithm is an extended version of the existing heuristic vehicle allocation algorithm named successive best insertion (SBI) [2]. The SBI is a semi-optimal vehicle allocation algorithm operating in realtime with limited computational resources. The optimization of the SBI is based on a cost function. We define new cost functions considering user satisfaction preference and use the SBI to allocate vehicles maximizing user satisfaction. We switch cost functions based on user satisfaction preference because individual users have different satisfaction preferences. In this paper, we define cost functions based on the user requirements of *quick arrival* and *the lowest cost*, which correspond to the user satisfaction preference of convenience and economy, respectively.

To confirm the effectiveness of our vehicle allocation algorithm, we conducted simulation experiments with requests extracted from a dataset of taxi trips on Manhattan Island in New York City. The experimental results reveal that our vehicle allocation algorithm successfully improved user satisfaction. Specifically, our main contributions are twofold:

- We present the design of a vehicle allocation algorithm with the consideration of user satisfaction preference. We use the existing SBI algorithm with newly defined cost functions to allocate vehicles maximizing user satisfaction.
- We show the improvement of user satisfaction compared to an existing vehicle allocation algorithm by up to 11% in simulation experiments using a dataset of taxi trips on Manhattan Island in New York City.

The remainder of the paper is organized as follows. Section 2 looks through related work on ride-sharing systems. We present our vehicle allocation algorithm in Sect. 3, followed by simulation evaluations in Sect. 4. Section 5 concludes the paper.

2 Related Work

Vehicle allocation in a ride-sharing services with sequentially generated requests is one of the Dial-a-Ride Problems (DARPs) [10]. We assume a dynamic situation where requests are sequentially generated. References [11, 12] pointed out that optimization in dynamic situations is more difficult than in static situations where requests are reserved in advance. To solve a DARP in real time, quasi-optimization is a practical approach. For this reason, algorithms that optimize in about polynomial time are widely used in actual ride-sharing services [13–15].

In the field of ride-sharing, pioneering work demonstrated the effectiveness of ride-sharing services compared to fixed-route buses while keeping the travel distance low [2, 3]. Based on the pioneering work, ride-sharing optimization is studied. Previous studies have investigated the optimization of vehicle routing [4, 5], vehicle allocation [6], vehicle location [7], and rational pricing [8, 9]. However, user satisfaction with ride-sharing services has not been well studied.

User satisfaction in ride-sharing services has primarily been defined in terms of arrival time [16, 17]. In these studies, the time difference between estimated and actual arrival due to the ride-sharing is considered to affect user satisfaction. User satisfaction, however, is difficult to model only using the arrival time. Reference [18] reported that user satisfaction in ride-sharing services is dependent not only on the arrival time, but also on many aspects including convenience and safety. There is a study defining user satisfaction based on both arrival time and fare [19]. Although these studies defined user satisfaction, no optimization was performed to maximize user satisfaction.

Levinger et al. proposed a vehicle allocation algorithm based on user satisfaction [20]. However, the proposed vehicle allocation algorithm assumes static requests from a single origin. In practical ride-sharing services, requests are sequentially generated by users with various origins and destinations. Beirigo et al. proposed the vehicle allocation algorithm that prioritizes and responds to requests according to the usage style selected by users [21]. There are three levels of usage styles: Business (1st), Standard (2nd), and Low-cost (3rd). Business (1st), which has the highest priority, is served by occupied vehicles and not shared rides, in exchange for a higher fee. Standard (2nd) and Low-cost (3rd) services are available at a lower cost than the Business (1st) services by sharing their rides. However, only three categories of usage patterns are set, making it difficult to respond to the various desires of users.

3 Vehicle Allocation Based on User Satisfaction

3.1 Use Case

In this paper, we assume the use case below.

1. A user requests a ride at any time to a ride-sharing management system. The request consists of the origin point, destination point, arrival deadline, and the number of passengers.
2. The ride-sharing management system allocates a vehicle to the request. A request acceptance signal including estimated boarding and arrival time is sent to the user. On allocation failure, a request rejection signal is sent to the user. A typi-

cal allocation failure occurs when the estimated arrival time exceeds the arrival deadline.
3. When the user receives an acceptance signal, the user boards the allocated vehicle at the origin point specified by the user. Immediately after the vehicle arrives at the destination point, the user gets off.
4. The allocated vehicle might carry another user on the way to the origin and destination points specified by the user. Therefore, the actual boarding and arrival time might be different from the estimated boarding and arrival time. Note that the requested information never changes during a ride.
5. The ride-sharing management system calculates the fare for the ride based on the ride time and an arrival delay from the estimated arrival time, which is charged to the user.
6. Users send feedback after using the ride-sharing service. In the feedback, the user explicitly selects from three options: "desire to spend cheaply", "desire to travel faster" or "satisfied with the service".

3.2 *Overview of Vehicle Allocation*

Figure 1 illustrates the overview of our vehicle allocation algorithm. Our key idea is to switch multiple cost functions based on an individual user satisfaction preference in a vehicle allocation optimization process. In this algorithm, we find the vehicle with minimum additional cost and allocate the request. The additional cost of each vehicle is calculated using the cost functions selected based on user satisfaction preference.

A user satisfaction preference is represented by a combination of convenience weight W_C and economy weight W_E, where $W_C + W_E = 1$. We assume that a user satisfaction preference dynamically changes. We therefore update the weights for

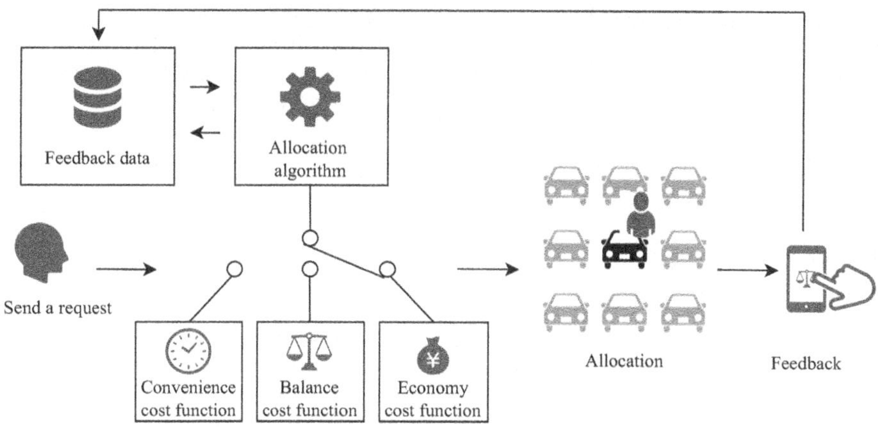

Fig. 1 Overview of vehicle allocation algorithm

each user based on feedback after the service use. In this paper, the ratio of W_C and W_E is assumed to be unchanged from the pre-defined ratio.

In this paper, we define three cost functions: convenience, economy, and balance cost functions. In the vehicle allocation process, we calculate the cost of vehicle allocation for each vehicle using the cost function depending on the weights of a new request. When $W_C > W_E$, we use the convenience cost function in the cost calculation. When $W_C < W_E$, the economy cost function is used. The balance cost function is used when $W_C = W_E$. Note that we always use the convenience cost function for vehicles with no request because the economy and balance cost functions incur a huge delay when there is no request other than a new request.

We do not limit the vehicle allocation optimization algorithm. In this paper, we use the successive best insertion (SBI) algorithm [2].

The following subsections describe the basics of the SBI algorithm, followed by the definition of the cost functions.

3.3 Successive Best Insertion (SBI) Algorithm

The SBI is a vehicle allocation algorithm for ride-sharing services [2], which calculates semi-optimal vehicle allocation with limited computational resources.

Algorithm 1 summarizes the process of the SBI algorithm. In Algorithm 1, *cost_func*(*route*) is the cost function over the given *route*, *time*(D) is the estimated time of arrival at destination D, and $\binom{x}{y}$ represents a binomial coefficient. The SBI is a two-step optimization process. In the first step, the minimum cost caused by the new request is calculated for each vehicle. L_ζ is a list of origins/destinations of vehicle ζ. We insert origin O_n and destination D_n of a new request to each position of L_ζ and create a list *routes*[ ] of new route candidates on lines 5–6. The number of the new route candidates is calculated by a binomial coefficient $\binom{l+2}{l}$, where l is the number of origins/destinations assigned to vehicle ζ. We change index number k within a range from 1 to $\binom{l+2}{l}$ to store each route candidate in *routes*[ ]. On lines 8–11, a cost for each route candidate is calculated. The cost of vehicle ζ is then calculated by finding the route candidate with a minimum cost on lines 12–13. In the second step, find the vehicle with the minimum cost and allocate the vehicle to the new request on lines 14–15.

3.4 Cost Functions

We define convenience, economy, and balance cost functions based on each of user's satisfaction preferences of convenience and economy. In ride-sharing services, a fare discount is applied when users allow shared rides, when shared rides occur, and when delays occur. The economy in this study is represented by reduced fares. We therefore define a new cost function to allocate vehicles prioritizing non-economic users, which

Algorithm 1: Successive Best Insertion (SBI)

Data: List Z of vehicles, list L_ζ of origins O_i and destinations D_j corresponding to requests assigned to vehicle ζ, origin O_n and destination D_n of new request, *deadline* of arrival time

Result: $v \Leftarrow$ vehicle with minimum cost, $L \Leftarrow$ new origin/destination list including O_n and D_n

all $cost[\,] = \infty$;
for ζ *in* Z **do**
  all $routes[\,]$ = empty list;
  $l = sizeof(L_\zeta)$;
  for k *in* $\binom{l+2}{2}$ **do**
    $routes[k] = L_\zeta$ with inserted O_n and D_n;
  all $cost\_r[\,] = \infty$;
  for *route in routes*[ ] **do**
    $cost\_r[route] = cost\_func(route)$;
    if $time(D_n) > deadline$ **then**
      $cost\_r[route] = \infty$;
  $L'_\zeta = \arg\min_{route} cost\_r[route]$;
  $cost[\zeta] = cost\_r[L'_\zeta]$;
$v = \arg\min_\zeta cost[\zeta]$;
$L = L'_v$;

promotes shared rides to users with preferences of economy. The convenience cost function C_C is defined with an estimated arrival time of a new request and arrival delay of the requests assigned to a vehicle, which is incurred by the new request. Let R be a set of requests assigned to a vehicle. Note that a new request is not included in R. The convenience cost function C_C is defined as:

$$C_C = t_{EA} - t_{EEA} + \sum_{r \in R} W_{C,r}\delta(r), \tag{1}$$

where t_{EA} is the absolute time of estimated arrival. We insert the origin and destination of the new request into each position of the vehicle's origin/destination list and calculate t_{EA} for each origin/destination set. t_{EEA} is the absolute time of estimated earliest arrival when a vehicle picks up a user immediately after the user's request and travels to the destination on the shortest route. We calculate t_{EEA} by adding the shortest travel time to the time that the user requests a ride. $\delta(r)$ is the delay incurred by the new request to request r, which is one of the requests already assigned to the vehicle. For each request r in R, we calculate the estimated arrival time for r when we insert the new request and subtract the new arrival time from the estimated arrival time without the new request, deriving $\delta(r)$.

The term $\sum_{r \in R} W_{C,r}\delta(r)$ in formula (1) is the sum of delays weighted by the convenience weight $W_{C,r}$. The sum is calculated over all the requests assigned to a vehicle. We can take user satisfaction preferences into account for the requests already assigned to a vehicle. When the user satisfaction preference ratio is $W_C : W_E = 1.0 : 0.0$, as an extreme example, the delay time incurred by the request is 100% included in the cost. When the user satisfaction preference ratio is $W_C : W_E = 0.0 : 1.0$, the delay time is excluded from the cost calculation. In short, a greater economy weight W_E incurs arrival delays.

For economy-oriented users, we define an economy cost function. The economy cost function C_E is defined as:

$$C_E = t_D - t_{EA} + \sum_{r \in R} W_{C,r}\delta(r), \tag{2}$$

where t_D is the arrival deadline for the new request. The economy cost function aims to deliver users a ride as cheaply as possible arriving before the arrival deadline. We therefore designed the economy cost function to be small when the time length of the shared ride is long.

For users who want a balanced response between convenience and economy, we define a balance cost function. The balance cost function C_B is defined as:

$$C_B = \left| \frac{1}{2}(t_D + t_{EEA}) - t_{EA} \right| + \sum_{r \in R} W_{C,r}\delta(r). \tag{3}$$

The balance cost function aims to balance arrival time and fare. By equalizing the weights of convenience and economy, $\sum_{r \in R} W_{Cr}\delta(r)$, i.e., the sum of the delay on requests assigned to a vehicle, is likely to be minimized. Compared to the convenience cost function C_C, the impact of the term $\sum_{r \in R} W_{Cr}\delta(r)$ is small in C_B. The balance cost function therefore minimizes the sum of the delay while prioritizing the minimum delays for convenience-oriented users.

4 Evaluation

To confirm the effectiveness of the proposed method, we conducted simulation experiments. We first evaluated economy user satisfaction against the user satisfaction preference weights. We also evaluated the impact of fare parameters on economy user satisfaction. User satisfaction with different weight distributions was evaluated because the number of users with specific user satisfaction preference weights affects the mean user satisfaction.

4.1 Experiment Environment

We used the Simulation of Urban MObility (SUMO) simulator. The SUMO is an open-source road traffic simulator, which is equipped with the Traffic Control Inter-

Fig. 2 Road network

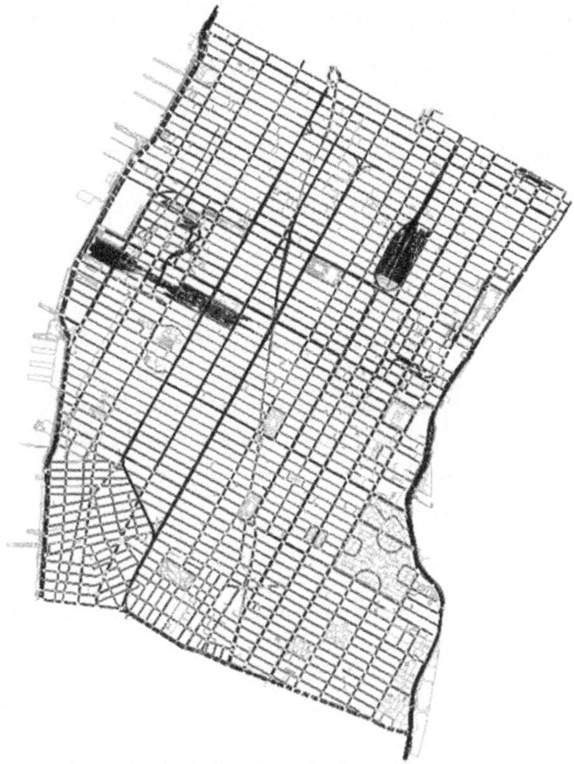

face (TraCI) allowing us to integrate external programs to control vehicle behavior [22]. We implemented a vehicle allocation algorithm in Python, which was integrated into the SUMO via the TraCI.

Figure 2 shows a road network used in the simulation. The road network is a part of Manhattan Island and covers an area of approximately 20 km^2, which was constructed using OpenStreetMap.[1] The numbers of nodes and edges, i.e., roads and intersections, are 20454 and 11133, respectively.

Requests were generated based on the taxi trip open data.[2] We randomly extracted requests from 09:00 am to 10:00 am on weekdays in May 2019. The number of extracted requests was determined as the mean daily number of requests from 09:00 am to 10:00 am on weekdays in May 2019. The request data contains boarding and arriving time, boarding and arriving locations, and the number of passengers. The boarding and arriving locations are recorded as an area ID,[3] which is uniquely given to each of the divided areas in Manhattan. In our experiment, we randomly chose edges in the boarding and arriving areas, respectively, as boarding and arriving edges.

[1] https://www.openstreetmap.org/.

[2] https://www1.nyc.gov/site/tlc/about/tlc-trip-record-data.page/.

[3] https://data.cityofnewyork.us/Transportation/NYC-Taxi-Zones/d3c5-ddgc/.

Table 1 Simulation parameters

Parameter	Value
# of requests	4456
# of person/request	1
# of vehicles	800, 1000, 1200
Seating capacity	4
Vehicle speed V [km/h]	30

Table 1 shows simulation parameters. A vehicle can carry up to 4 persons in addition to a driver. The simulation started at time $t = 0$ and stopped at $t = 10000$ seconds. Requests were generated between $t = 400$ and $t = 4000$ second.

To avoid impossible request generation, the arrival deadline t_{D} for each request was determined as:

$$t_{\mathrm{D}} = t_{\mathrm{O}} + m\frac{d}{V} + T_{\mathrm{P}}, \tag{4}$$

where t_{O} is the time when the request occurs, d is the distance between origin and destination points of the request, m is a margin coefficient, and T_{P} is a constant value corresponding to the mean time required for pick-up. The margin coefficient m is a simulation parameter for calculating the rational deadline.

In this experiment, T_{P} and m were set to 600 s and 1.5, respectively. Each vehicle was put at a random point at $t = 0$ second. An empty vehicle kept the current position until a new request was assigned.

To demonstrate relative performance, we compared the user satisfaction of the following two methods.

(P) Proposed: The proposed method presented in Sect. 3. Cost functions are switched based on the users' preference.

(O) Original: The original SBI, which minimizes the number of request rejections. The SBI uses the cost function C_{O} defined as:

$$C_{\mathrm{O}} = t_{\mathrm{EA}} - t_{\mathrm{EEA}} + \sum_{r \in R} \delta(r), \tag{5}$$

which is designed to minimize the arrival delay caused by new requests.

4.2 User Satisfaction Metric

To validate the effectiveness of our proposed vehicle allocation, we evaluated user satisfaction using a user satisfaction metric. A user satisfaction preference is dependent on users. We define a user satisfaction metric S as:

$$S = W_C S_C + W_E S_E, \tag{6}$$

where S_C and S_E are convenience user satisfaction and economy user satisfaction metrics, respectively. The user satisfaction metric S takes a value between 0 and 1; a higher value indicates higher user satisfaction. As described in Sect. 3.2, W_C and W_E are given by the user. S_C and S_E are defined as follows.

Convenience user satisfaction metric S_C: For convenience-oriented users, we use the convenience user satisfaction metric S_C defined as the ratio of the actual remaining time to the estimated remaining time until the arrival deadline:

$$S_C = \frac{t_D - t_A}{t_D - t_{EA}}, \tag{7}$$

where t_A is the actual drop-off time. S_C takes a value between 0 and 1 because actual arrival is never before the estimated arrival and never after the deadline.

Economy user satisfaction metric S_E: For economy-oriented users, we use the economy user satisfaction metric S_E defined as the ratio of discounted fare to the original fare:

$$S_E = \max\left\{0, 1 - \frac{F_B + dF_U - (t_A - t_O)F_D}{F_B + dF_U}\right\}, \tag{8}$$

where F_B is a fixed base fare, F_U is the fare per unit distance, and F_D is the fare discount per unit time delay. We emphasize that F_B, F_U, and F_D have different dimensions: F_B has the dimension of currency, while the dimensions of F_U and F_D are currency per distance and currency per time, respectively.

S_E takes a value between 0 and 1. Formula (8) is based on the evaluation formula used in the Japanese ride-sharing service named Smart Access Vehicle Service (SAVS) [13]. Formula (8) consists of terms corresponding to a maximum fare and a rebate charge. The maximum fare is calculated from the sum of F_U and F_B multi plied by d, where d is the distance between the origin and destination of a request. The rebate charge is calculated by multiplying time required for arrival, which is calculated by subtracting t_O from t_A, multiplied by F_D. The later the arrival time t_A becomes due to sharing or traffic congestion, the lower the fare becomes.

For example, if F_D is increased relative to F_U, a small delay can significantly increase user satisfaction. However, if F_D is excessively large compared to F_U, it becomes difficult to make it profitable. In this formula, the value of user satisfaction varies greatly depending on the value of F_U and F_D. In this experiment, we set $F_B = 300$ JPY, $F_U = 0.4$ JPY$/m$, and $F_D = 0.4$ JPY$/s$, referring to Ref. [23].

4.3 User Satisfaction Against the User Satisfaction Preference Weights

Figure 3 shows the mean user satisfaction. The horizontal and vertical axes indicate the number of vehicles and user satisfaction, respectively. We can fairly compare user satisfaction because all requests were accepted by both methods. Figure 3 shows that the proposed method has higher user satisfaction than the original method for all cases of the number of vehicles. We can confirm that the smaller the number of vehicles, the higher user satisfaction with the proposed method. When the number of vehicles is 800, user satisfaction is improved by 11% compared to the original method.

Figures 4 and 5 show the user satisfaction as a function of the user satisfaction preference weight. Figures 4 and 5 indicate the following:

- The proposed method tends to have a higher convenience user satisfaction for users who placed more importance on convenience, and a lower convenience user satisfaction for users who did not place importance on convenience. In particular, there is a significant decrease in convenience user satisfaction for users who do not value convenience for the parameters with 800 and 1000 vehicles.
- The proposed method shows a drop in convenience user satisfaction only when the number of vehicles is 1200 and the weight of convenience is 0.1. Because there were sufficient vehicles in response to requests, the algorithm incurred not so many shared rides.
- Economy user satisfaction is higher for the proposed method when the weight of the economy is 0.5 or higher.
- The proposed method has the highest economy user satisfaction when the number of vehicles is 800. Because the supply of vehicles in response to requests is lower than that of the other number of vehicles, the algorithm incurred many shared rides, which resulted in lower fares.

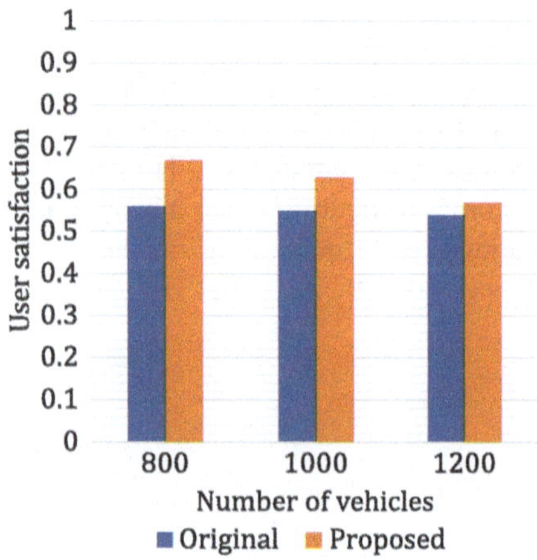

Fig. 3 Mean user satisfaction

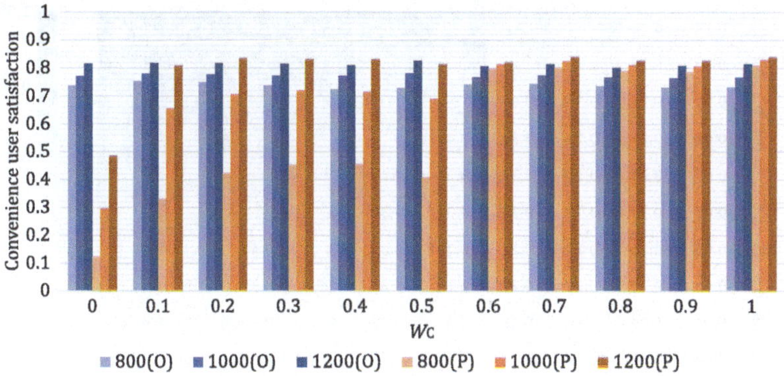

Fig. 4 Mean convenience user satisfaction by weight. For each vehicle number of 800, 1000, and 1200. (O) and (P) in the legend represent the original and proposed methods

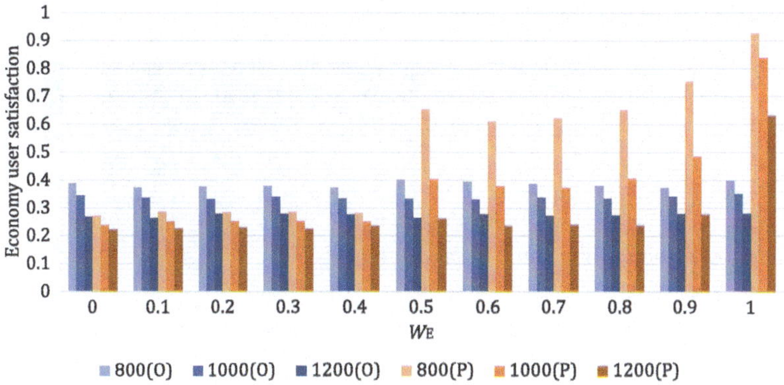

Fig. 5 Mean economy user satisfaction by weight. For each vehicle number of 800, 1000, and 1200. (O) and (P) in the legend represent the original and proposed methods

4.4 Economy User Satisfaction Against Different Fare Parameters

Next, we conducted further analysis of the economy user satisfaction. In Sect. 4.3, the economy user satisfaction was evaluated only with the parameters of $F_B = 300$ JPY, $F_U = 0.4$ JPY$/m$, and $F_D = 0.4$ JPY$/s$. The parameters F_U and F_D have a significant impact on the economy user satisfaction. Therefore, we analyzed the change in the economy user satisfaction when the parameters F_U and F_D were changed.

Figures 6 and 7 show the economy user satisfaction of the original and proposed methods when F_U and F_D are changed from 0.1 to 1.0. The horizontal and vertical axes indicate parameters F_D and F_U, respectively. Figures 6 and 7 indicate the following:

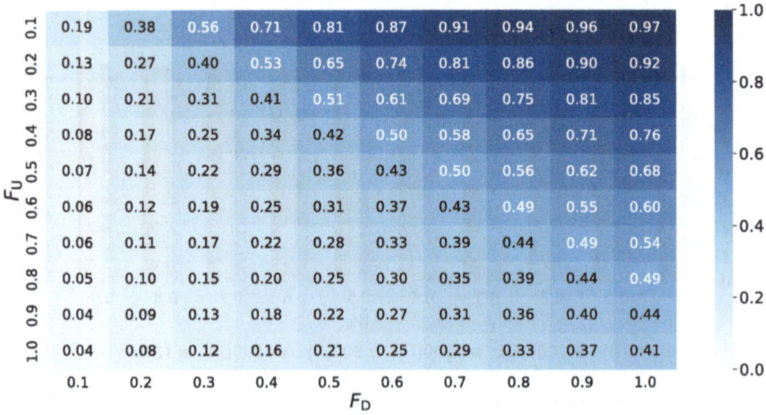

Fig. 6 Economy user satisfaction for each parameter of the original method. F_{U} is the fare per unit distance, and F_{D} is the fare discount per unit time delay

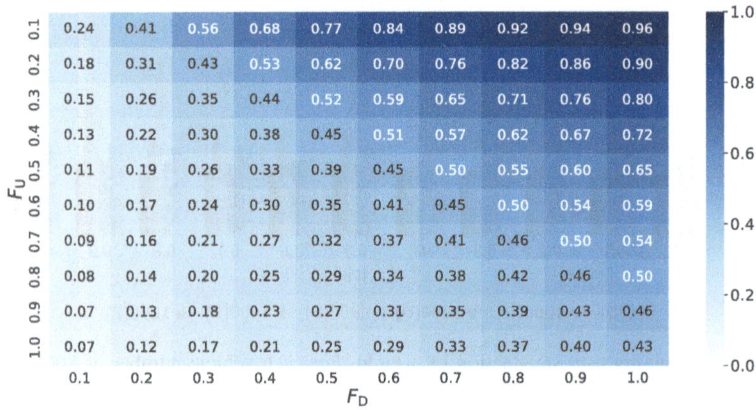

Fig. 7 Economy user satisfaction for each parameter of the proposed method. F_{U} is the fare per unit distance, and F_{D} is the fare discount per unit delay time

- Both methods show higher values of economy user satisfaction when F_{D} is large compared to F_{U}. Economy user satisfaction is higher when F_{D} is large compared to F_{U}. This is natural because a small delay significantly reduces the fare.
- When F_{D} is large compared to F_{U}, the original method tends to have higher economy user satisfaction, while the proposed method has higher economy user satisfaction when F_{D} is small compared to F_{U}. However, the configurations such that F_{D} is large compared to F_{U} are impractical in terms of profitability because the fare discount is too large.

4.5 User Satisfaction Against Different Weight Distribution

To validate the effectiveness of the proposed method, we conducted an experiment with requests for the distribution of various user satisfaction preferences. In Sect. 4.3 experiment, the distribution of user satisfaction preference weights is determined randomly. However, the actual distribution of user satisfaction preference weights is likely to be diverse. The diversity of the distribution may vary based on a variety of factors, including hours of operation, the population of the service area, the distribution of users' annual income, and the distribution of user age groups.

In our experiment, we assumed six patterns of the distribution of user satisfaction preference weights. Figure 8 shows the distribution of the weights used in our experiment. The horizontal and vertical axes indicate the convenience weights and the

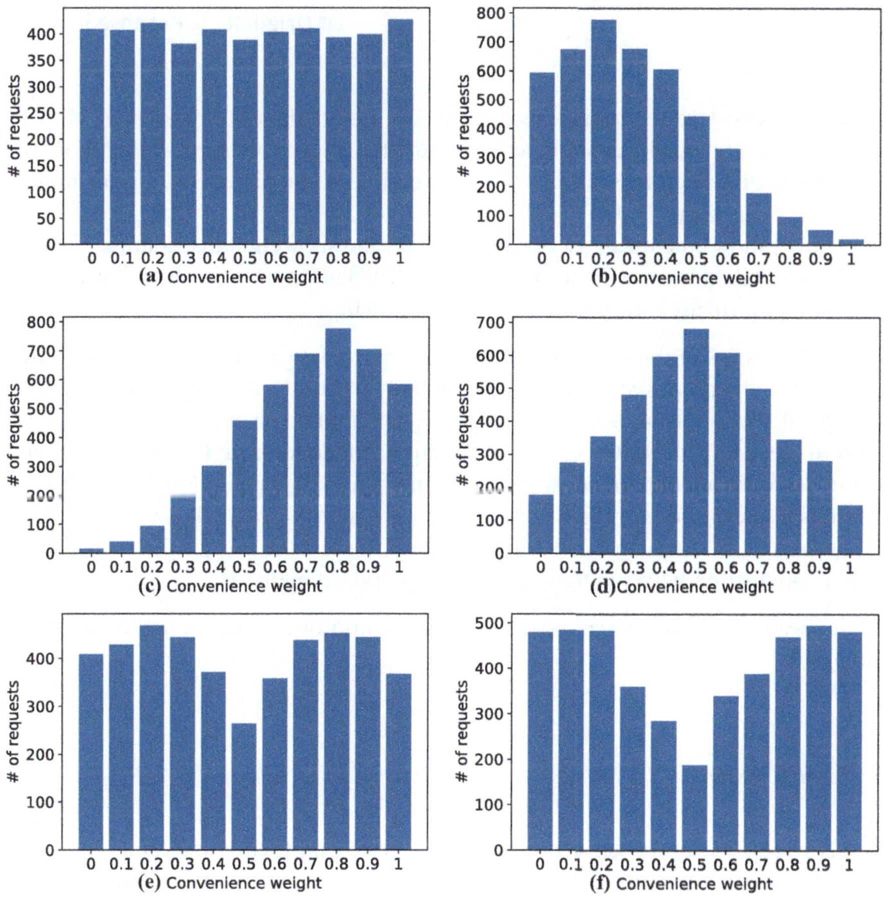

Fig. 8 Six patterns of the distribution of user satisfaction preference weights. The weight distribution represents the distribution of users' requests corresponding to each weight

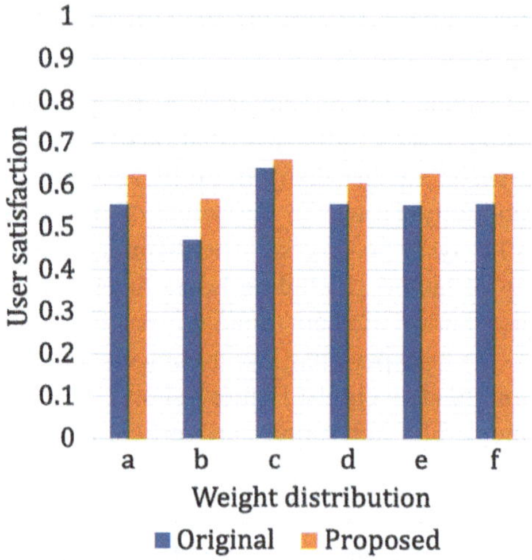

Fig. 9 Mean user satisfaction for each weight distribution (number of vehicles = 1000)

number of generated requests, respectively. The weight distribution shown in Fig. 8a is the one used in the aforementioned experiment. Other experimental conditions are the same as described in Sect. 4.1. In this experiment, the parameter for the number of vehicles was set to 1000.

Figure 9 shows the mean user satisfaction for the six weight distribution patterns. The horizontal and vertical axes indicate the distribution of requests and user satisfaction, respectively. Figure 9 indicates the following:

- For all patterns of the user satisfaction preference weight distributions, user satisfaction of the proposed method is higher than that of the original method.
- User satisfaction is higher for distributions with more convenience-oriented users. From the results in Sect. 4.4, user satisfaction tended to be higher for convenience user satisfaction than for economy user satisfaction. It is natural that user satisfaction is higher for distributions with more convenience-oriented users.
- The larger the number of economy-oriented users is, the greater the difference in user satisfaction between the original and proposed methods.

These results show that the proposed method can improve user satisfaction in various situations.

5 Conclusion

In this paper, we proposed a ride-sharing service employing a vehicle allocation algorithm using three cost functions to improve user satisfaction. The cost function used in the vehicle allocation is switched based on the individual user satisfaction preference. We conducted simulation experiments and demonstrated the performance of the proposed switching vehicle allocation algorithm improved user satisfaction by up to 11% while maintaining vehicle allocation performance.

In this paper, we assumed that user satisfaction preference is represented by a combination of convenience and economy. However, user satisfaction can also be considered from preferences other than convenience and economy. For example, ride comfort and the outside view can be also considered to become a factor of user satisfaction. In our future work, we plan to design a new cost function to support other user satisfaction preferences.

Acknowledgements This work was supported in part by JSPS KAKENHI Grant Number JP20K11793.

References

1. Yaraghi N, Ravi S (2017) The current and future state of the sharing economy. Brookings India IMPACT, vol 032017
2. Noda I, Ohta M, Shinoda K et al (2003) Evaluation of usability of dial-a-ride systems by social simulation. In: Multi-agent-based simulation III (MABS2003), LNCS, pp 167–181
3. Ma S, Zheng Y, Wolfson O (2015) Real-time city-scale taxi ridesharing. IEEE Trans Knowl Data Eng 27(7):1782–1795 Jul
4. Lin Q, Deng L, Sun J et al (2018) Optimal demand-aware ride-sharing routing. In: IEEE INFOCOM, pp 2699–2707
5. Alonso-Mora J, Wallar A, Rus D (2017) Predictive routing for autonomous mobility-on-demand systems with ride-sharing. In: IEEE/RSJ international conference on intelligent robots and systems (IROS), pp 3583–3590
6. Miao F, Han S, Lin S et al (2016) Taxi dispatch with real-time sensing data in metropolitan areas: a receding horizon control approach. IEEE Trans Autom Sci Eng 13(2):463–478 Apr
7. Iglesias R, Rossi F, Wang K et al (2018) Data-driven model predictive control of autonomous mobility-on-demand systems. In: 2018 IEEE international conference on robotics automation (ICRA), pp 6019–6025
8. Asghari M, Shahabi C (2017) An on-line truthful and individually rational pricing mechanism for ride-sharing. In: ACM SIGSPATIAL, pp 1–10
9. Asghari M, Shahabi C (2018) ADAPT-pricing: a dynamic and predictive technique for pricing to maximize revenue in ridesharing platforms. In: ACM SIGSPATIAL, pp 189–198
10. Cordeau J-F, Laporte G (2007) The dial-a-ride problem: models and algorithms. Ann Oper Res 153(1):29–46 Sep
11. Agatz N, Erera A, Savelsbergh M et al (2012) Optimization for dynamic ride-sharing: a review. Eur J Oper Res 223(2):295–303
12. Furuhata M, Dessouky M, OrdÓnez F et al (2013) Ridesharing: the state-of-the-art and future directions. Transp Res Part B: Methodol **57**:28–46
13. Nakashima H, Sano S, Hirata K et al (2016) One cycle of smart access vehicle service development. In: Serviceology for designing the future. ICServ 2014, pp 247–262
14. Bischoff J, Maciejewski M, Nagel K (2017) City-wide shared taxis: a simulation study in Berlin. In: Proceedings 2017 IEEE ITSC, pp 275–280
15. Iwata S, Suzuki K (2019) Two way allocation methods in SAVS for large depopulated area. In: Asia pacific symposium on intelligent and evolutionary systems (IES2019), pp 33–45
16. Lin Y, Li W, Qiu F et al (2012) Research on optimization of vehicle routing problem for ride-sharing taxi. Proc - Soc Behav Sci 43:494–502
17. Yang J, Li J, Chen Y et al (2013) Multi-objective distribution model and algorithm for online shopping express logistics. J Comput 8(10):2558–2564 Oct

18. Jahan M (2019) Factors affecting customer satisfaction of the ride-sharing industry in Bangladesh. Bus Ethics and Leadership 3(4):74–80 Dec
19. Atasoy B, Ikeda T, Song X et al (2015) The concept and impact analysis of a flexible mobility on demand system. Transp Res Part C: Emerg Technol 56:373–392 Jul
20. Levinger C, Hazon N, Azaria A (2020) Human satisfaction as the ultimate goal in ridesharing. Futur Gener Comput Syst 112:176–184 Nov
21. Beirigo BA, Negenborn RR, Alonso-Mora J et al (2022) A business class for autonomous mobility-on-demand: modeling service quality contracts in dynamic ridesharing systems. Transp Res Part C: Emerg Technol 136:103520 Mar
22. Lopez PA, Behrisch M, Bieker-Walz L et al Microscopic traffic simulation using SUMO. In: Proceedings of 2018 IEEE ITSC, pp 2575–2582
23. Noda I (2020) Case study of price system design for taxi sharing by co-evolutional process between service corporations and users. In: IPSJ, vol 2020-ICS-198, Issue 16, pp 1–7. (in Japanese)

A City Airspace Testbed for Drone Networks in Future Smart Cities

Taku Yamazaki, Sumiko Miyata, and Takumi Miyoshi

Abstract Drones are the key technology for supporting future smart city services from the airspace. To validate the feasibility of city services, service developers must implement city airspace the system in real-world settings such as urban areas. However, the experimenters cannot avoid various limitations and legal regulations. In this chapter, we propose a city airspace testbed that can validate the feasibility of drone networks and services in real-world settings. Our testbed has four key features: modularity, portability, safety, and agility for proof-of-concept in urban areas. Our testbed comprises lightweight self-standing smart poles as the key modules that act as hovering drones rather than physical ones. Finally, we demonstrated the three fundamental field experiment scenarios using our testbed. First, we demonstrated the emulation of multiple hovering drones. Second, we conducted a wireless communication experiment to emulate the communication between drones and ground devices. Third, we conducted a monitoring experiment by capturing 4K 360-degree photos and videos from a drone and streaming the video from the drone to the ground user.

Acronyms

AERPAW	Aerial Experimentation and Research Platform for Advanced Wireless
AGV	Automated Guided Vehicle
AI	Artificial Intelligence

T. Yamazaki (✉) · S. Miyata · T. Miyoshi
Shibaura Institute of Technology, Tokyo, Japan
e-mail: taku@shibaura-it.ac.jp

S. Miyata
e-mail: sumiko@shibaura-it.ac.jp

T. Miyoshi
e-mail: miyoshi@shibaura-it.ac.jp

R. Shinkuma et al. (eds.), *Advances in Engineering and Information Science Toward Smart City and Beyond*, Engineering Cyber-Physical Systems and Critical Infrastructures 5, https://doi.org/10.1007/978-3-031-29301-6_7

COSMOS	Cloud Enhanced Open Software Defined Mobile Wireless Testbed for City-Scale Deployment
CaaS	City as a Service
GENI	Global Environment for Networking Innovation
IoT	Internet of Things
LiDAR	Laser Imaging, Detection and Ranging
ORBIT	Open-Access Research Testbed for Next-Generation Wireless Networks
SSID	Service Set Identifier
UAV	Unmanned Aerial Vehicle

Glossary

City Airspace Low altitude airspace over a city where drones can fly.

City Airspace Testbed A testbed that realize to validate the feasibility of drone networks and services in real-world settings such as in a city where the experimenters want to deploy these services.

Data Line Cables that transfer data between equipment on the ground and a smart pole for remote control and data analysis.

Module Equipment that experimenters deploy a device into an experiment field such as on the ground and a smart pole to realize a function such as sensing, computing, networking, and supplying power.

Power Line Cables that supply electric power from ground power sources for equipment on a smart pole.

Self-Standing Pole A pole that can stay upright using mounted rotors by generating wind power against the tilt of the pole.

Self-Standing Smart Pole A self-standing pole that experimenters can attach various modules as payload. The pole virtually acts as a hovering drone in real city fields.

1 Introduction

Smart cities [1–3] are expected to obtain the provision of more intelligent and efficient city services by enabling information and communication technologies to improve people's quality of life while reducing the environmental burden. Particularly, the city as a service (CaaS) concept [4, 5] facilitates people's ability to choose and receive necessary city services in their daily lives. For example, the city of Espoo has been working on implementing the CaaS concept [6–8].

These smart city services are based on artificial intelligence (AI) technologies for predictions and suggestions to aid decision-making. Therefore, enormous amounts of various real-time sensing data must be comprehensively collected and utilized depending on the use cases. The new digital twin paradigm virtually represents the real world in real time for the comprehensive circulation and multidimensional utilization of this big data. Therefore, digital twin is compatible with AI-enabled smart city services because smart city services are tightly coupled with the physical city space. That is, real-time sensing and monitoring of the actual city space to support digital twin are required to obtain future smart cities.

Drones play essential roles in the urban airspace to gather sensing data [9–11], logistics [12–14], surveillance [15, 16], search and rescue [17, 18], and disaster prevention [19, 20]. In addition, various AI technologies are expected to obtain these services [21, 22]. Therefore, to deploy such AI-enabled smart city services in city airspace, developers must conduct field experiments in the city and analyze the effectiveness of the developed service. However, some limitations exist in implementing drone services in actual city airspace. First, they must comply with many legal regulations for flying drones [23–27]. Second, the experimenters must guarantee the safety of drones in flight to avoid accident risks such as falling drones. Third, the current commercial drone's flight time is limited to approximately 1 h even in a large quadcopter [28]. In addition, all the experiment equipment must be mounted to the flying drone. However, the payload of the drone has a limit. These limitations are unavoidable if the experimenters reproduce the service in actual city environments using current commercial or developed drones. However, they affect experiments on future city services based on the breakthrough of drone technologies. Therefore, the testbed aimed at innovative city services is required to avoid the aforementioned limitations and legal regulations in urban areas. Consequently, it will obtain the rapid and easy deployment of the developed system and service.

Various experiment frameworks and platforms have been studied and developed from the perspective of proof-of-concept for information networking [29–32], city services [33–40], and drone services [41–46]. However, the service developers should conduct field trials to validate the feasibility of their systems because the city services are strongly related to the city and people. However, a fixed or indoor testbed cannot satisfy these requirements. Although several outdoor testbeds in cities are available, they do not primarily focus on actual flights of drones, and actual field experiments are subject to some limitations and legal regulations, as aforementioned.

We propose a city airspace testbed as a proof-of-concept of drone networks and services in future smart cities. Our testbed focuses on validating the feasibility of drone services in actual cities. Our testbed satisfies four key requirements: modularity, portability, safety, and agility. Our testbed primarily comprises self-standing smart poles that act as hovering drones instead of physical ones. Our self-standing smart pole retains its posture upright to use wind power generated by multiple rotors when it detects its tilt. In addition, the smart poles are easy to carry in experiment fields because they are divisible and lightweight. Experimenters can easily deploy them in experiment fields because the poles are only put in experiment positions and do not require actual flight in city airspace. Hence, our testbed ensures safety while an experiment reduces accident risks caused by out-of-control or battery exhaustion. In addition, the experimenters can easily mount various experiment modules, including their system, on the smart poles. The experimenters can rapidly validate the service feasibility in various fields where they must implement the developed system.

The remainder of this chapter is organized as follows. Section 2 surveys the related work on testbeds for services. Section 3 introduces our drone service testbed for future smart cities. Section 4 demonstrates testbed use cases and their results. Finally, Sect. 5 concludes this chapter and suggests some future work.

2 Related Work

We present various studies for the experiment framework and platform related to developing information networking, drone services, and city services.

First, we introduce the citywide and nationwide network testbeds [29–32]. ORBIT testbed [29, 30] provided a realistic evaluation platform for network protocols and applications. This testbed allows using an indoor wireless grid with the radio propagation emulator and an outdoor network for real-world field trials. The ORBIT testbed is helpful for testing various services. GENI [31, 32] obtained a networking testbed nationwide. This testbed can validate the feasibility of the services on wide-scale networking and distributed systems. However, these testbeds primarily focus on the experiments of networking services and protocols using citywide and nationwide fixed facilities. Therefore, they might not be suitable for executing services associated with the actual individual city environments.

We introduce the citywide and nationwide city service testbeds [33–40]. LOG-a-TEC [33, 34] is a set of testbeds that realizes outdoor sensing platform. The LOG-a-TEC primarily addresses the internet of things (IoT) services using low-power wireless communication technologies. Although the testbed focuses on the comprehensive and heterogeneous IoT environment, it currently does not focus on airspace like drones. COSMOS [35, 36] realized the city-scale wireless testbed for real-world experiments. The testbed has been deployed in an actual city. Therefore, COSMOS covers the smart city services; experimenters demonstrated a smart intersection service [37]. However, this platform might not focus on drone services in smart cities. AERPAW [38] realized a real testbed for aerial experimentation and research platform of drones. However, AERPAW mainly focuses on the actual performance analysis of low-level communication protocols and algorithms because it is not deployed in rural areas but in metropolitan areas. Therefore, it is not easy to validate the feasibility of smart city services for metropolitan areas. Toyota Woven City [39] also realized the citywide testing space in Susono City, Japan. Although Toyota Woven City includes the usage of drones, such services have not been experimented with thus far. Stockholm UAM Lab. [40] realized the citywide testing space in ICT Arena, Kista Science City, Sweden. In addition, UAM Lab includes a drone in the actual city. It is an ideal and comprehensive approach to realize the experiment environment; however, the current status is unknown.

In addition, various indoor drone testbeds were developed [41–46]. Amador et al. [41] developed indoor small drone testbed for the target tracking. Khan et al. [42] developed an indoor drone experiment environment using object detection. Palacious et al. [43] developed a testing platform for services of heterogeneous unmanned vehicles. Afanasov et al. [44] developed the indoor testbed focusing on high-level drone applications. Ridolfi et al. [45] focused on warehouse automation using unmanned aerial vehicles (UAVs) and automated guided vehicles (AGVs). Qin et al. [46] developed the experiment environment for cooperative distributed sensing in a small room. All the testbeds realized the testing environment focusing on use cases. However, because of their design concepts, they are difficult to apply to other scenarios and services. In addition, real outdoor environments differ from indoor environments in some ways.

In summary, the developers should conduct field trials to validate the feasibility of the developed systems and services because the city services are strongly related to the city and its people. However, a fixed or indoor testbed cannot satisfy these requirements. By contrast, several outdoor testbeds are available. However, actual field experiments are subject to some limitations and legal regulations [23–27].

Therefore, several points must be satisfied to confirm the feasibility of smart city services using drones in urban areas. First, the testbed must be adapted to the various use cases in smart cities. Second, the testbed can be used in urban areas where the experimenters must deploy their services. Third, the testbed must be safe to deploy in these areas, even if there is a flow of people. Finally, the testbed should be easy and agile to deploy the city services.

3 City Airspace Testbed for Future Smart Cities

This section presents a city airspace testbed that is a proof-of-concept platform for future drone networks and services in smart cities. First, we define and introduce four important concepts of our city airspace testbed to satisfy the requirements of experimenters: modularity, portability, safety, and agility. As the key module of our testbed, we use a self-standing smart pole that virtually acts as a hovering drone. The experimenters can validate the feasibility of city services covered from the ground to the low-altitude airspace to deploy a whole testbed composed of smart poles and their systems in real fields.

Figure 1 shows the architecture of our city airspace testbed and the tentative procedure for using it. Before conducting an actual field experiment, the experimenters must make a detailed design to clarify (1) the location and schedule, (2) the experiment model and required equipment, and (3) the intended experiment situations and deployment design of equipment. Then, they can choose any location for the experiment because of the portability and safety of our testbed, as described later. In addition, they can determine the whole system model for the experiment and the required equipment as modules based on the modularity of our testbed, including both sides of the experimenter and testbed. Next, they determine the deployment

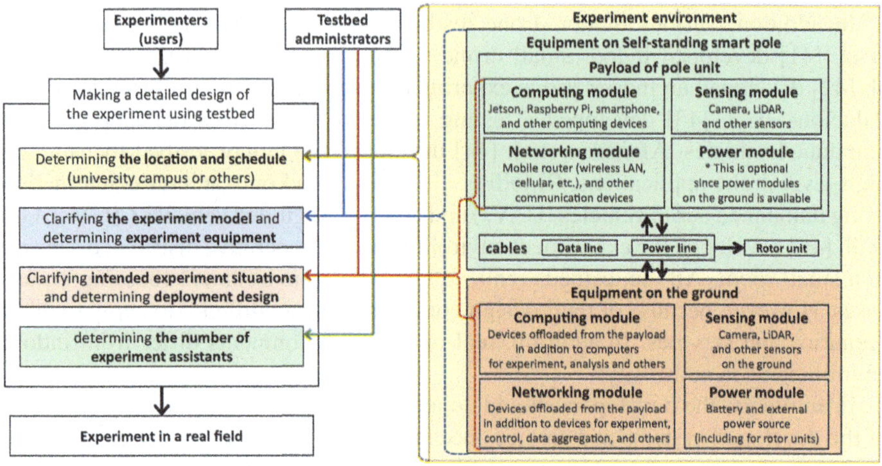

Fig. 1 Architecture of our city airspace testbed and the tentative procedure for using it

design of equipment based on the intended experiment situations. Then, they consider which modules should be mounted on a smart pole because our testbed can offload modules to the ground. This step is important to simplify the experiment to focus on the proof-of-concept. Finally, experimenters determine the number of assistants for smart poles and conduct the real-field experiment.

3.1 Key Concepts

Our testbed has four important concepts to satisfy the requirements of experimenters compared to existing testbeds as explained in Sect. 2: (1) modularity, (2) portability, (3) safety, and (4) agility.

Modularity Our testbed is composed of various modules; hence, experimenters can select and use only the experiment modules they require. Our testbed primarily consists of self-standing smart poles and equipment related to the individual field experiment. In the minimum case, the experimenters carry the smart poles to an experiment field, and they can confirm whether drones can be deployed in the field virtually. The experimenters can conduct field experiments by adding the necessary equipment as a module. As a result, they can focus on the essential points for their proof-of-concept without influence from the outside elements by maintaining the system scale of proof-of-concept to the minimum necessary without any additional development burdens. Note that our testbed includes computing modules (Jetson, Raspberry Pi, smartphones, and other computing devices), sensor modules (camera, LiDAR, atmospheric sensor, anemometer, and other sensors), networking modules (wireless LAN router, cellular communication router, and other communication devices), and power modules (DC battery, USB battery, and others) in addition to the smart pole as the key module of our testbed. In addition, modules can be offloaded to the ground depending on assumptions in the experiment. For example, the experimenters can put high-performance computers for heavy computational tasks and batteries on the ground and connect them to devices on the smart pole via wired lines to provide computational resources and power sources.

Portability Our testbed is fully portable because it primarily comprises the easy-to-carry lightweight smart poles and the minimum equipment for the system because of the modularity of the testbed. Therefore, the experimenters can easily carry the entire testbed, including the developed system, to the experiment field. That is, they can directly deploy the whole system developed in the laboratory anywhere and validate its feasibility in the actual field.

Safety Our testbed is relatively safe to demonstrate the experiment systems in real-world settings because smart poles are used as hovering drones instead of real drones. In contrast to physical drones in flight, no risks of out-of-control and falling drones are involved, even if problems occur. Generally, physical drones in flight always contain substantial potential risks of falling from the airspace because of out-of-control situations because of battery exhaustion, sudden wind, and control errors. A large drone weighs nearly 1 kg without the payload, even in the case of general consumer drones; hence, it is impossible to avoid injury if fallen drones hit people. In addition, serious injury may occur if the rotors are hit. Therefore, the experimenters must control the flow of people or forbid the intrusion of unrelated people into the drones' flight range. However, it is a considerable limitation for validating the feasibility of city services because smart city services also are strongly related to living people. Consequently, direct deployment of physical drones without enough pre-experiments to ensure safety is difficult, even if the actual flight of physical drones is not essential for proof-of-concept.

Agility Our testbed realizes agility for the proof-of-concept considering the concepts mentioned above. First, the experimenters deploy the minimum developed system for the proof-of-concept in real fields based on our modular-based testbed. Because the self-standing smart poles act as hovering drones in city airspace, they can avoid the limitations and legal regulations for the experiment in urban areas. Finally, they can carry the testbed with the proof-of-concept system from their laboratories because of the portability. Hence, they directly deploy the entire system to the real smart city fields. Therefore, experimenters do not require expending considerable effort derived from the deployment in real fields.

3.2 Self-standing Smart Pole as a Hovering Drone

We present the fundamental behavior of our self-standing smart pole and the specifications of the smart pole for use as the key module of the testbed.

To realize our testbed, we must find alternatives to physical drones for reproducing the behavior of drones. Our testbed uses a self-standing pole. The inverted pendulum [47] is a well-established approach to self-standing. However, the bottom part of the pole must be adjusted to stand upright. That is, deploying such a pole to the actual field is difficult because it must reserve enough space for the margin. In addition, the authors could not find sufficiently long poles to satisfy our requirements for emulating flying drones. Another approach is using wind power generated by multiple rotors to maintain the pole's upright posture. If the pole detects its tilt, it immediately adjusts itself to the correct angle with the ground. In contrast to the approach mentioned above, it can be fixed at a single point on the ground, although the top part of the pole moves, which can save ground space for the margin. Although the commercial one based on the wind power [48] was developed by a company [49], its length is approximately 1.5 m and short for our use case.

We adopted a long self-standing pole under development to express the hovering drone in collaboration with the company [49]. Figure 2 shows a picture of the self-standing pole, and Fig. 3 shows its demonstration of self-standing. To attach various equipment to the self-standing pole, we utilize it as a self-standing smart

Fig. 2 Self-standing smart pole that is divided into a pole unit and rotor units

Fig. 3 Demonstration of our self-standing smart pole in a real field

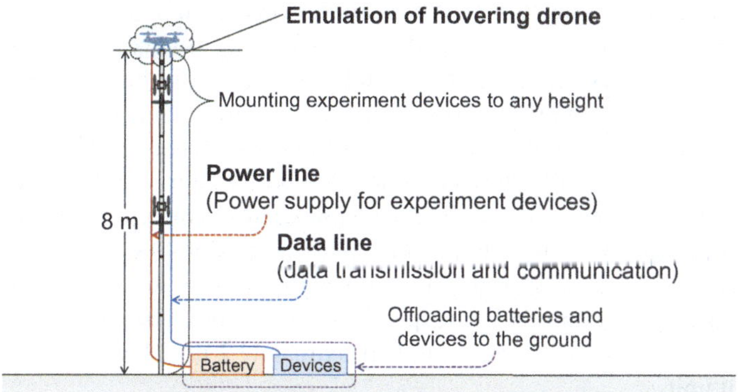

Fig. 4 Overview of the self-standing smart pole

pole, which has the communication and computational resources in our testbed. We defined specifications for our self-standing smart pole, which will be used as the key module.

Figure 4 depicts the overview of the self-standing smart pole. Smart poles stay upright using mounted rotors by generating wind power against the tilt of the pole. Therefore, they maintain their posture even if the poles are tilted because of the wind effect. In addition, the smart pole can mount experimental equipment, such as the drone's payload, at any height. Table 1 indicates the current specification of

Table 1 Current specification of self-standing smart pole

Parameter	Value
Total weight	3890 g (1 pole unit and 2 rotor units)
Posture control	wind power generated by rotor units
Weight of pole unit	2400 g
Minimum length of pole unit	1915 mm
Maximum length of pole unit	8000 mm (current version)
Weight of roter unit	745 g
Number of rotors of rotor unit	4 rotors
Diameter of rotors	256 mm
Maximum payload for additional equipments	1000 g (depending on the shape and balance of mounted equipments)
Power line options for additional equipments	DC cable, USB, PoE, or any power cables
Data line options for additional equipments	USB, wired LAN, or any data cables

our self-standing smart pole. According to the fundamental specifications, the smart pole consists of a pole unit and two rotor units. The pole unit comprises 8 short poles with approximately 1.5 m length. Therefore, experimenters can shorten them to a minimum of approximately 1.9 m for storing or transporting them somewhere. The maximum length of the pole unit is 8 m in the current implementation. Note that the developed pole unit can be extended to over 12 m in maximum. However, we currently limit the maximum length to 8 m because of the stability of self-standing. Therefore, the experimenters can create emulation environments where the drones fly up to 8 m. Note that each pole can express multiple drones to mount some equipment at different height positions. The rotor unit consists of 4 rotors (256 mm in diameter), which are pairs of 2 rotors, to generate wind power in any direction on the horizontal axis for posture control. In addition, the smart pole is 3890 g and can be divided into pole and rotor units. Each unit is lightweight because the weights of pole and rotor units are 2400 g and 745 g, respectively. Therefore, our smart pole is highly portable because it can be shortened to approximately 2 m or less and separated into lightweight units.

The self-standing smart pole can mount data and power lines for functional extensions, as shown in Fig. 4. The data line provides the communication function from the ground for controlling the mounted devices or measuring and analyzing data if necessary. In some cases, future services require computation resources for machine learning technology. Assuming such a situation, we can generate a situation wherein the drone has sufficient computing resources in real time by placing high-performance devices on the ground and connecting them. The power line provides energy for the rotors and mounted devices from the power supply at the ground. In contrast to real drones, the smart pole can increase the payload available by offloading the battery on the ground. In addition, the smart pole can stand semi-permanently. In the experiment, the maximum flight time of commodity drones is up to approximately 1 h if

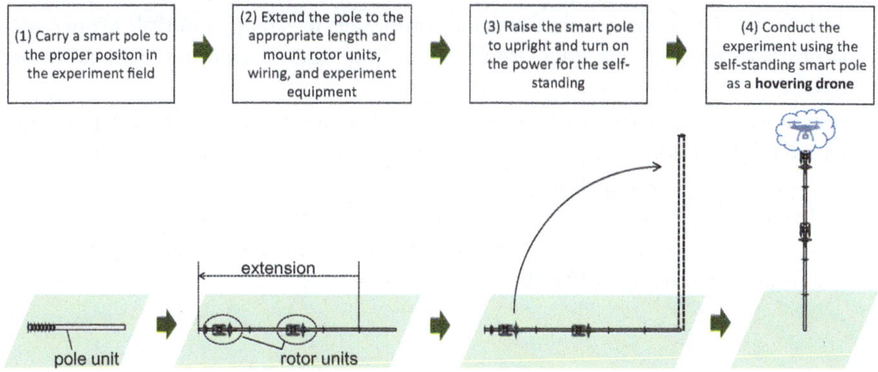

Fig. 5 Procedure of activating self-standing smart pole

we use real drones for demonstrations. Our smart pole can maintain virtual hovering using external power on the ground. Therefore, the battery capacity and its weight have no limitations. The experimenters must supply power to rotor units via any one of three ways depending on experiment situations: (1) built-in battery on the rotor unit, (2) external battery put on the ground, or (2) external power source on the ground. We adopted the second choice because it can easily extend the lifetime by replacing the empty battery with a full one on the fly, and the users can easily experiment anywhere without using external power outlets.

In our testbed, activating standing smart poles requires four steps, as illustrated in Fig. 5: (1) carry the smart pole to the experiment position, (2) extend the pole and mount equipment to the pole, (3) raise the smart pole to be upright and turn on the power to start the self-standing, and (4) start to conduct the main experiment. Note that we require an assistant who always stays at the root of the smart pole to ensure safe self-standing in the current implementation. In addition, another assistant is required to raise the smart pole to avoid sudden risks of falling poles.

4 Field Experiments

In this section, we demonstrate the usage of our proposed testbed. In the first phase of the testbed development, we conducted all experiments in the multipurpose court of the Omiya Campus, Shibaura Institute of Technology, Japan, for safety. During the experiments, assistant students waited at each smart pole underneath one-by-one to avoid sudden accidents. We then conducted the following three experiments assuming the use of drones for city services.

Experiment 1 We experimented with emulating multiple hovering drones. This experiment confirms the smart poles' stability and feasibility of emulating multiple flying drones using self-standing smart poles. This experiment reveals the feasibility of the demonstrations of various drone services using multiple drones such as protocol analysis for inter-drone networking and cooperative control among drones.

Experiment 2 We experimented with emulating air-to-air communication in an actual field. In addition to confirming the connectivity of hovering drones, we analyzed the communication quality.

Experiment 3 We monitored and captured video from the viewpoint of a hovering drone. The experiment confirms the feasibility of using a self-standing smart pole as a drone for sensing, surveillance, and monitoring.

4.1 Experiment 1: Emulating Multiple Hovering Drones using Self-standing Smart Poles

4.1.1 Experiment Setup

As the first step to confirm the feasibility of our testbed in real-world settings, we demonstrated the ability to create multiple self-standing smart poles simultaneously, assuming multiple hovering drones. Figure 6 illustrates the assumptions of the experiment. In this experiment, we adopted four self-standing smart poles. The smart poles mount only the weight as the payload for the stability of self-standing. This experiment was conducted with a calm wind during the day to clarify the primitive characteristics.

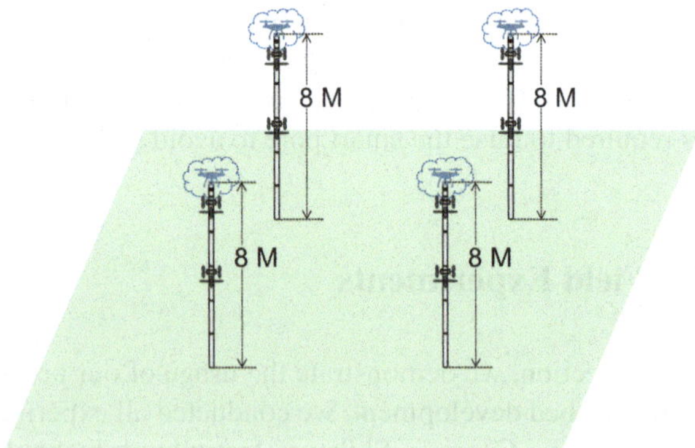

Fig. 6 Experiment 1: Scenario of emulating multiple hovering drones

Fig. 7 Experiment 1: Demonstration of emulating multiple hovering drones with a height of 8 m

4.1.2 Experiment Results

Figure 7 shows the demonstration of multiple virtually hovering drones in the actual field. The wind power generated by their rotors maintained the stable self-standing of all the smart poles. Therefore, the smart poles, the key module in our testbed, were designed to stand simultaneously for reproducing multiple flying drones. In addition, despite being tilted at times in relatively stronger wind, all of the smart poles did not fall during the experiment. The smart poles never become out-of-control, such as with faulty rotor behaviors. The smart poles slowly tilted and persistently endured the wind load even if an unexpectedly strong wind blew. The smart poles could cancel this unexpected strong wind by quickly changing the wind power with their rotors. The experimenters can use the testbed in environments where the wind is blowing, including an unexpectedly strong wind.

4.2 Experiment 2: Inter-drone Communication and Its Quality

4.2.1 Experiment Setup

As the second step, we conducted an experiment demonstrating wireless communication between drones and ground devices using our testbed. Figure 8 shows the

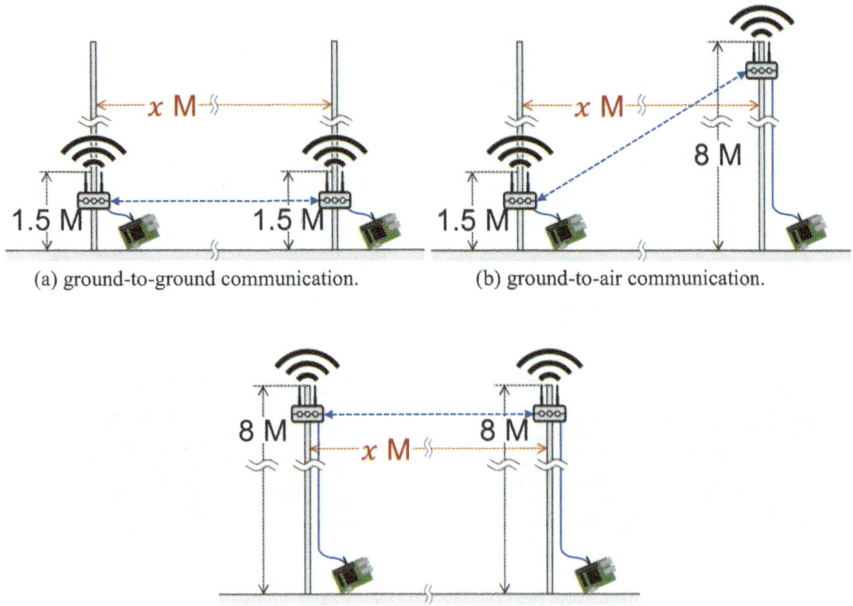

(a) ground-to-ground communication.

(b) ground-to-air communication.

(c) air-to-air communication.

Fig. 8 Experiment 2: Scenarios

individual scenarios in the experiment. We assumed three types of scenarios: (1) ground-to-ground, (2) ground-to-air, and (3) air-to-air communications. The scenarios focus on analyzing the difference between the propagation environments of the ground and the airspace. Ground-to-ground communication is assumed to be between the fixed facilities or users' devices on the ground. Therefore, this scenario confirmed the fundamental property of wireless communication for comparison with the following scenarios. Ground-to-air communication is assumed that a fixed facility or user on the ground communicates with a flying drone. In this scenario, we assumed use cases such as remote sensing using a drone and network access through a flying access point as the temporary network infrastructure in disaster-hit areas. Air-to-air communication is assumed to be the use case of information sharing or cooperation between drones, such as federated learning in the city airspace and cooperative sensing. In addition, we confirmed the communication quality in these situations as the use case of our testbed. Therefore, this experiment primarily focuses on the applicability of our testbed for confirming these scenarios.

We used two wireless mobile routers, GL.iNet GL-AXT1800 [50] as the wireless communication devices, and they were mounted at a height of 8 m on the smart pole as a payload for the hovering drone. As the device is on the ground, we fixed the device at the height of 1.5 m from the ground. GL-AXT1800 has a Qualcomm IPQ6000, which can use IEEE 802.11ax (Wi-Fi6) with 245 g weight. We only used the 2.4 GHz band without using the 5 GHz band, and the communication speed was up to 573 Mbps because the 5 GHz band is prohibited for outdoor use in the airspace [52]. Assuming that the embedded device serves as the computational resource and is virtually mounted to the drones, we adopted two Raspberry Pi 4 Model B with 8 GB RAM [51]. In this experiment, we directly put the Raspberry Pi on the ground. The Raspberry Pi was connected to the GL-AXT1800 through a LAN cable, and two GL-AXT1800 were connected through IEEE 802.11ax. In this experiment, mobile batteries placed on the ground powered all the experiment devices using a USB-C cable. In addition, GL-AXT1800 has external omnidirectional antennas that can adjust their angles. We set the antennas upright in the same manner as in the case of using them on the ground. Note that the antenna characteristics in the vertical direction might affect the experiment results.

By using these devices, we generated the traffic for the throughput measurement. We used iPerf (version 3.9) [53] as the network traffic generator and the throughput measurement software. iPerf was installed in Raspberry Pi and generated the TCP traffic from a client to a server. We adopted TCP CUBIC [54] as the congestion control algorithm, and the TCP maximum segment size (MSS) was set to 1448 bytes. We configured two Raspberry Pi for the client and the server of iPerf. We generated TCP traffic in 60 s from the client to the server for the throughput measurement. We conducted the measurement five times and collected the experiment results. We note that the devices were remotely controlled by a laptop computer connected to the wireless LAN. Hence, the results include the effect of the remote control. In the case of ground-to-air communication, we confirmed the performances in both directions, that is, for ground-to-air and air-to-ground communications. We changed the measurement distance x between the wireless devices from 10 to 50 m.

Note that other courts, woods, sidewalks, and roads primarily surround the experiment field. Although there are no tall buildings around the experiment field, our results might include a few effects from them such as reflections. In addition, radio interference from existing wireless LAN systems might affect the real-field experiments. Therefore, we observed wireless LAN beacon frames, which advertise the service set identifier (SSID) of access points, at the experiment field. As a result, beacon frames were observed, including wireless LANs of our university, and their signal strengths were approximately lower than −70 dBm. Therefore, although our results might include the effect from the other deployed wireless LAN systems, there are few effects because the signal strengths are weak, and there are no office buildings around the experiment field.

4.2.2 Experiment Result

Table 2 shows the experiment results. The results of 10 m distance of all scenarios are the highest average throughput. Particularly, the average throughput of air-to-air communication is the highest, irrespective of the distance. By contrast, ground-to-air and air-to-ground communications have relatively smaller throughput than the others. In particular, the throughput of ground-to-air and air-to-ground communications largely varies based on the distance x. Although the results include the effect of actual

Table 2 Experiment 2: Throughput analysis

Scenario	Distance *x* (m)	Average (Mbps)	Median (Mbps)	Maximum (Mbps)	Minimum	SD
Ground-to-Ground	10	128.2	130	155	14.2 Mbps	15.0
	30	114.1	115	159	10.6 Mbps	16.6
	50	102.2	104	120	30.6 Mbps	12.3
Ground-to-Air	10	125	126	158	16.8 Mbps	15.9
	30	123.4	124	156	0 bps	19.0
	50	69.7	70.8	96.1	5.58 Mbps	13.0
Air-to-Ground	10	125.0	126	158	16.8 Mbps	15.9
	30	114.5	115	145	31 Mbps	13.6
	50	79.2	80.6	99.4	23 Mbps	11.1
Air-to-Air	10	296.8	303	362	72.8 Mbps	39.7
	30	177.0	179	232	35.6 Mbps	26.0
	50	156.4	160	188	39.4 Mbps	19.4

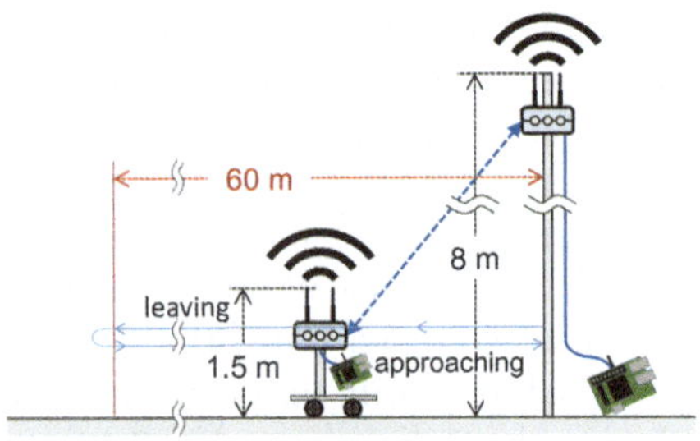

(a) Scenario.

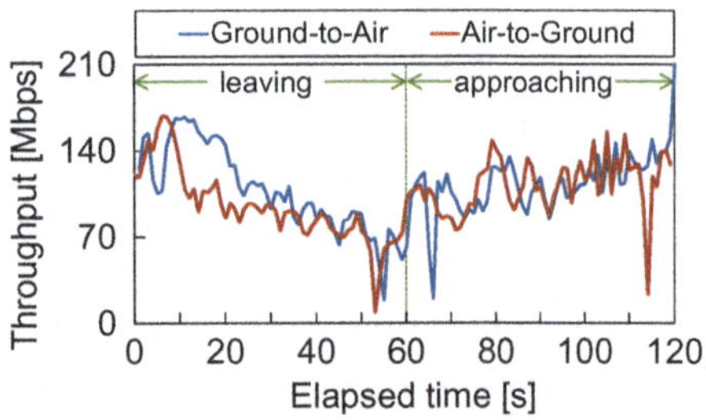

(b) Transition of the throughput.

Fig. 9 Experiment 2: Walking scenario and result

device-to-device distance, the decreased throughput ratio is large compared to the ground-to-ground communication case. Although one of the reasons seems to be the effect of ground reflection, the results might include the effect of the antenna pattern of the adopted device and wireless interferences among other wireless LAN systems. As a result, the testbed can quickly confirm and analyze these three-dimensional effects in real fields without flying drones.

Additionally, assuming the two situations of accessing a flying access point and receiving a stream from a flying drone, we demonstrated the throughput measurements while walking below and around the drone. Figure 9a depicts the scenario of the experiment. GL-AXT1800 and Raspberry Pi with mobile batteries were fixed on a cart as the experimenter's device, and the experimenter walked while pushing the cart. The experimenter started walking from the smart pole's root to the opposite side of the multipurpose court with a walking speed of 1 m/s. The distance is approximately 60 m. Subsequently, the experimenter returned to the smart pole's root with the same walking conditions.

Figure 9b shows the throughput transition until walking. This result indicates that the throughput gradually decreased and increased depending on the distance between the hovering drone and the experimenter, as mentioned previously. In addition, the measured throughput largely fluctuated until walking. Therefore, our testbed could reproduce such a natural variation of results in the mixed situation of the ground and airspace. In addition, the service developer who requires high and stable throughput must handle this behavior by protocol and service design.

4.3 Experiment 3: Taking Video from Hovering Drones

4.3.1 Experiment Setup

In the third step, we were challenged to capture 360-degree photos and videos in addition to the video streaming from a smart pole that acts as a hovering drone. We adopted RICOH THETA V [55] as the 360-degree camera device mounted on the smart pole. THETA V can capture a 4K (3840 $\times$ 1920 pixels) 29.97 fps 360-degree video and stream the video to the other device in real-time. THETA V has IEEE 802.11n as the wireless medium for remote control and real-time 360-degree video streaming. In addition, we additionally equipped a 3D microphone RICOH TA-1 to THETA V to record the 3D environmental sound in addition to the video. THETA V and TA-1 weighed approximately 184 g (121 g and 63 g, respectively).

We assumed the situation in which the experimenter remotely controls the 360-degree camera mounted on the flying drone. Therefore, we confirmed two types of situations: capturing photos and videos via the remote control and viewing the streaming video via the smartphone as the experimenter's client, as illustrated in Fig. 10. We used a single smart pole in this experiment and mounted the THETA V at its 8 m height. THETA V includes the battery; thus, we did not use the power cable for THETA V in the experiment. We set up the THETA V as the wireless access point mode. We connected THETA V to the client software installed on a user's smartphone. We confirmed the quality of the captured photos and videos for the feasibility of the drone's emulations of remote sensing, monitoring, and surveillance from city airspace.

4.3.2 Experiment Result

Figures 11 and 12 show the 360-degree screenshot as well as the corrected and cropped screenshots from the captured video by the 360-degree camera. First, we could successfully mount the 360-degree camera on the smart pole and reproduce to capture the clear video from the flying drone without any problems. Therefore, we can easily demonstrate the abovementioned services using the self-standing smart pole as the hovering drone. However, the captured video contained some vibration caused by rotors and the self-standing control of the smart pole while canceling the

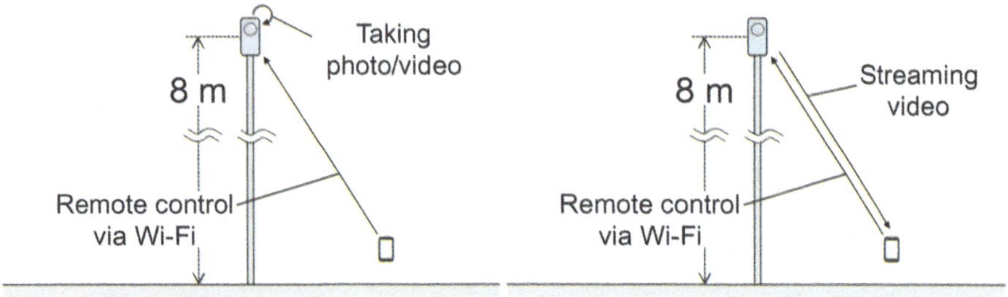

Fig. 10 Experiment 3: Scenarios

Fig. 11 Experiment 3: Screenshot of capturing 360-degree video being captured from the viewpoint of the hovering drone at 8 m height

(a) Campus side. (b) Court side (opposite side).

Fig. 12 Experiment 3: Screenshots of two corrected viewpoints from the captured video in Fig. 11

wind effect from various directions. As mentioned previously, the self-standing smart poles maintain their posture using the wind power generated by rotors. Therefore, they cause the vibration from the rotors and shake from their posture control. By contrast, the wind also strongly affects physical drones. That is, there is a possibility that our testbed can obtain similar results to the environments where experimenters use physical drones affected by real wind.

Additionally, we recorded the sound in parallel to capturing a video. The sound of the captured video includes the environmental sounds such as the experimenters' voices. However, some constant noises were caused by the rotors of the smart pole and natural wind. Therefore, the experimenters might handle noises, such as noise cancellation, when collecting small environmental sounds, although the effect from the rotors is the same as that of physical drones.

In addition, we could see the 360-degree streaming video while capturing the video from around the smart pole. The experimenter could successfully see the video streaming through the smartphone application, although the video sometimes includes lag, noise, and stops. The adopted camera device can communicate with others through IEEE 802.11n. Therefore, demonstrations of real-time monitoring via streaming video to the user's device can also be confirmed through our testbed.

5 Conclusions

We proposed a city airspace testbed as a proof-of-concept platform for drone networks and services in future smart cities. Our testbed primarily comprises self-standing smart poles as hovering drones to substitute for real drones. Self-standing smart poles can maintain an upright posture using the wind power generated by their rotors. Using smart poles as hovering drones, we achieved safe environments for deploying developed city services in the real world. Our testbed has four key features for satisfying the experimenters' requirements: (1) modularity, (2) portability, (3) safety, and (4) agility. On the basis of key features, our testbed provided an environment where experimenters could rapidly demonstrate the drone-based city services to safely deploy the developed system without significant modifications using the smart poles as hovering drones rather than real drones.

In addition, we conducted three experiments to clarify the effectiveness of the smart pole as the key module of our testbed. The flight experiment confirmed that the experimenters could demonstrate smart city services, including multiple drones, through our testbed. The communication experiment confirmed that the experimenters could demonstrate and analyze the performance of city services from the viewpoint of communication quality. The monitoring experiment confirmed that the experimenters could demonstrate capturing video from low-altitude airspace. Therefore, our testbed can satisfy the requirements of experimenters seeking to validate the feasibility of innovative smart city services in realistic environments.

In future work, we plan to address three challenges: (1) proof-of-concept for actual drone services for smart cities using our testbed in an actual city environment, (2) more detailed analysis and development of the behavior of smart poles as hovering drones, and (3) improvement of the openness and accessibility of our testbed for external experimenters.

During the experiments, our smart poles were stably standing in calm environments. However, the stability of poles in windy environments has scope for improvement. We particularly confirmed that the self-standing stability of poles strongly depends on the weight of the mounted payload and the wind speed of the experiment environment. In this study, we validated that the poles functioned with hundreds of grams of payload and additional cables attached to the poles. We will further investigate the relationship between the weight of the payload, wind power, and stability.

We plan to control the tilt of self-standing smart poles to reproduce drones' more natural vibration and shaking behavior. In this study, the smart poles only adjusted their tilt to stay vertical to the ground using the wind power of rotors. However, if the smart poles generate more wind power than the necessary power for the adjustment in any direction, they can intentionally tilt in addition to vibrate and shake. Therefore, we will analyze the adjustment behavior of poles. In addition, we will consider the algorithm for generating this realistic behavior of physical drones.

In this study, we privately attempted to use our testbed as a test flight with internal and a few external experimenters. Therefore, after finishing the test trials of our testbed, we should make the testbed resources available to external users as experimenters to maximize the significance of our testbed. Therefore, we plan to create concrete usage procedures and guidelines for accessing our testbed resources.

Acknowledgements These research results were obtained from the commissioned research (No. 05601) by National Institute of Information and Communications Technology (NICT), Japan. We would like to express our special thanks to KUUSATSU GIKEN Inc., which gave us considerable support for developing the self-standing smart pole for our testbed, and Tomoki Nishizawa, who gave us helpful support for the real-field experiments.

References

1. Kirimtat A et al (2020) Future trends and current state of smart city concepts: a survey. IEEE Access 86448–86467. https://doi.org/10.1109/ACCESS.2020.2992441
2. Du R et al (2019) The sensable city: a survey on the deployment and management for smart city monitoring. IEEE Commun Surveys Tuts 1530–1560. https://doi.org/10.1109/COMST.2018.2881008
3. Camero A, Alba E (2019) Smart city and information technology: a review. Cities 84–94. https://doi.org/10.1016/j.cities.2019.04.014
4. Xu H, Geng X (2020) People-centric service intelligence for smart cities. Smart Cities 135–152. https://doi.org/10.3390/smartcities2020010
5. Klassen G, Buske M (2017) City as a service and city on-demand - new concepts for intelligent urban development. Digital Marketplaces Unleashed 795–807. https://doi.org/10.1007/978-3-662-49275-8_70
6. Anttiroiko AV (2016) City-as-a-platform: the rise of participatory innovation platforms in finnish cities. Sustainability 8:922. https://doi.org/10.3390/su8090922
7. Anttiroiko AV, Sahamies K (2022) Designing city service ecosystems: the case of the city of Espoo in the capital region of Finland. In: Distributed, ambient and pervasive interactions.

Smart environments, ecosystems, and cities. HCII 2022, pp 139–157. https://doi.org/10.1007/978-3-031-05463-1_10
8. City as a Service|City of Espoo. https://www.espoo.fi/en/city-espoo/innovative-espoo/city-service. Accessed 1 Sept 2022
9. Yanmaz E et al (2018) Drone networks: communications, coordination, and sensing. Ad Hoc Netw 1–15. https://doi.org/10.1016/j.adhoc.2017.09.001
10. Noor NM, Abdullah A, Hashim M (2018) Remote sensing UAV/drones and its applications for urban areas: a review. In: IOP conference series: earth and environmental science, pp 1–8. https://doi.org/10.1088/1755-1315/169/1/012003
11. Nex F, Remondino F (2014) UAV for 3D mapping applications: a review. Appl Geom 1–15. https://doi.org/10.1007/s12518-013-0120-x
12. Kückelhaus M, Niezgoda D, Unmanned aerial vehicles in logistics. https://www.dhl.com/discover/content/dam/dhl/downloads/interim/full/dhl-trend-report-uav.pdf. Accessed 1 Sept 2022
13. Rejeb A et al (2021) Drones for supply chain management and logistics: a review and research agenda. Int J Logist 1–24. https://doi.org/10.1080/13675567.2021.1981273
14. Gupta A et al (2021) Advances of UAVs toward future transportation: the state-of-the-art, challenges, and opportunities. Future Trans 326–350. https://doi.org/10.3390/futuretransp1020019
15. Sugano R et al (2022) Watch from sky: machine-learning-based multi-UAV network for predictive police surveillance, pp 1–6. arXiv:2203.02892. https://doi.org/10.48550/arXiv.2203.02892
16. Dilshad N et al (2020) Applications and challenges in video surveillance via drone: a brief survey. In: 2020 International Conference on Information and Communication Technology Convergence (2020) https://doi.org/10.1109/ICTC49870.2020.9289536
17. Mishra, B., et al.: Drone-surveillance for search and rescue in natural disaster. Computer Commun 1–10. https://doi.org/10.1016/j.comcom.2020.03.012
18. Albanese A (2022) SARDO: an automated search-and-rescue drone-based solution for victims localization. IEEE Trans Mob Comput 3312–3325. https://doi.org/10.1109/TMC.2021.3051273
19. Mohd Daud SMS et al (2022) Applications of drone in disaster management: a scoping review. Sci & Justice 30–42. https://doi.org/10.1016/j.scijus.2021.11.002
20. Restas A (2015) Drone applications for supporting disaster management. World J Eng Technol 316–321. https://doi.org/10.4236/wjet.2015.33C047
21. Azar AT et al (2021) Drone deep reinforcement learning: a review electronics, pp 1–30. https://doi.org/10.3390/electronics10090999
22. Osco PL et al (2021) A review on deep learning in UAV remote sensing. Int J Appl Earth Observ Geoinform 1–21. https://doi.org/10.1016/j.jag.2021.102456
23. Lee D, Hess D, Heldeweg MA (2022) Safety and privacy regulations for unmanned aerial vehicles: a multiple comparative analysis. Technol Soc 1–10. https://doi.org/10.1016/j.techsoc.2022.102079
24. Civil Aviation Bureau, Ministry of Land, Infrastructure, Transport and Tourism. https://www.mlit.go.jp/en/koku/index.html. Accessed 16 Dec 2022
25. Certificated remote pilots including commercial operators, Federal Aviation Administration. https://www.faa.gov/uas/commercial_operators. Accessed 16 Dec 2022
26. Overview: flying drones and model aircraft. UK Civil Aviation Authority. https://register-drones.caa.co.uk. Accessed 16 Dec 2022
27. Civil drones (unmanned aircraft), EASA. https://www.easa.europa.eu/en/domains/civil-drones. Accessed 16 Dec 2022
28. Matrice 300 RTK, Da-Jiang Innovations Science and Technology Co., Ltd. https://www.dji.com/jp/matrice-300. Accessed 16 Dec 2022
29. ORBIT Testbed - WINLAB, Rutgers University. https://winlab.rutgers.edu/research-projects/orbit-testbed. Accessed 6 Aug 2022
30. Raychaudhuri D et al (2005) Overview of the ORBIT radio grid testbed for evaluation of next-generation wireless network protocols. In: IEEE wireless communications and networking conference, pp 1664–1669. https://doi.org/10.1109/WCNC.2005.1424763

31. GENI https://www.geni.net/geni-partners/federated-testbeds. Accessed 1 Sept 2022
32. Berman M et al (2013) GENI: a federated testbed for innovative network experiments. Computer Netw, pp 5–23. https://doi.org/10.1016/j.bjp.2013.12.037
33. An architecture for fully reconfigurable plug-and-play wireless sensor network testbed. In: Proceeding of the 2015 IEEE global communications conference, pp 1–7. https://doi.org/10.1109/GLOCOM.2015.7417564
34. LOG-a-TEC by SensorLab. https://log-a-tec.eu. Accessed 1 Sept 2022
35. Raychaudhuri D et al (2020) Challenge: COSMOS: a city-scale programmable testbed for experimentation with advanced wireless. In: Proceedings of the 26th annual international conference on mobile computing and networking, pp 1–13. https://doi.org/10.1145/3372224.3380891
36. COSMOS - COSMOS Testbed Main Site. https://www.cosmos-lab.org. Accessed 1 Sept 2022
37. Kostić Z et al (2022) Smart city intersections: intelligence nodes for future metropolises, pp 1–8. https://doi.org/10.48550/arXiv.2205.01686
38. AERPAW - Aerial Experimentation and Research Platform for Advanced Wireless. https://aerpaw.org. Accessed 20 Sept 2022
39. Toyota Woven City. https://www.woven-city.global. Accessed 20 Sept 2022
40. Stockholm UAM Lab. http://uamlab.se. Accessed 20 Sept 2022
41. Amador JM et al (2019) An agile low-cost testbed for multi-drone target tracking. In: 2019 workshop on research, education and development of unmanned aerial systems, pp 344–351. https://doi.org/10.1109/REDUAS47371.2019.8999718
42. Khan M et al (2016) Simulating drone-be-gone: agile low-cost cyber-physical UAV testbed. In: Proceedings of the 2016 international conference on autonomous agents & multiagent systems, pp 1491–1492
43. Palacios FM et al (2017) Test bed for applications of heterogeneous unmanned vehicles. Int J Adv Robotic Syst 1–14. https://doi.org/10.1177/1729881416687111
44. Afanasov M et al (2019) FlyZone: a testbed for experimenting with aerial drone applications. In: Proceedings of the 17th annual international conference on mobile systems, applications, and services, pp 67–78. https://doi.org/10.1145/3307334.3326106
45. Ridolfi M et al (2019) Testbed for warehouse automation experiments using mobile AGVs and drones. In: IEEE conference on computer communications workshops, pp 919–920. https://doi.org/10.1109/INFCOMW.2019.8845218
46. Qin C et al (2002) 3, 2, 1, Drones Go! a testbed to take off UAV swarm intelligence for distributed sensing, pp 1–12. https://doi.org/10.48550/arXiv.2208.05914
47. Boubaker O (2013) The inverted pendulum benchmark in nonlinear control theory: a survey. Int J Adv Robot Syst 1–9. https://doi.org/10.5772/55058
48. TB-005PRO, Kuusatsu Giken Inc. https://taorenbo.com. Accessed 1 Sept 2022
49. Kuusatsu Giken Inc. https://www.multicopter.co.jp. Accessed 1 Sept 2022
50. GL.iNet GL-AXT1800. https://www.gl-inet.com/products/gl-axt1800. Accessed 1 Sept 2022
51. Raspberry Pi. https://www.raspberrypi.org. Accessed 1 Sept 2022
52. Radio Equipment for Unmanned Aircraft (UA), MIC the radio use website, Ministry of Internal Affairs and Communications. https://www.tele.soumu.go.jp/e/sys/others/drone/index.htm. Accessed 1 Sept 2022
53. iPerf, https://iperf.fr. Accessed 1 Sept 2022
54. Ha S, Rhee I, Xu L (2008) CUBIC: a new TCP-friendly high-speed TCP variant ACM SIGOPS. Oper Syst Rev 64–74. https://doi.org/10.1145/1400097.1400105
55. Ricoh Theta V, https://theta360.com/ja/about/theta/v.html. Accessed 1 Sept 2022

A Feasibility Study of Tethered Autonomous Moving Cells for Smart City

Kaori Ota, Hirofumi Watanabe, and Yu Nakayama

Abstract There will be more and more connected devices in the future smart cities. To deal with the spatio-temporal fluctuations of the traffic demand to/from these mobile devices, a moving mobile network has been investigated. Among them, a tethered flying drone is becoming a promising option for next generation mobile communications. However, to the best of our knowledge, tethered ground/floating nodes have not been employed for moving cells. Therefore, in this paper we propose a concept of tethered autonomous moving cells for smart cities. The goal of the tethered moving cells is to provide stable and high data rate communication via optical fiber cables for unstable demands. The tethered cables also provide long battery lifetime. We also propose the slackness control algorithm to optimally handle the optical cable. The feasibility of the proposed idea is demonstrated with the experimental results using a ground robot and a floating node.

Acronyms

C-RAN	Centralized Radio Access Network.
CU	Central Unit.
IoT	Internet of Things.
M2M	Machine to Machine.
mmWave	millimeter-Wave.

K. Ota (✉)
Department of Computer and Information Sciences, Tokyo University of Agriculture and Technology, 2-24-16 Nakacho, Koganei-shi, Tokyo 184-8588, Japan
e-mail: kaori.o@ieee.org

H. Watanabe
SoftBank Corporation, Tokyo Portcity Takeshiba Office Tower 27F, 1-7-1 Kaigan, Minato-ku, Tokyo 105-7529, Japan
e-mail: hirofumi01.watanabe@g.softbank.co.jp

Y. Nakayama
Institute of Engineering, Tokyo University of Agriculture and Technology, 2-24-16 Nakacho, Koganei-shi, Tokyo 184-8588, Japan
e-mail: yu.nakayama@ynlb.org

R. Shinkuma et al. (eds.), *Advances in Engineering and Information Science Toward Smart City and Beyond*, Engineering Cyber-Physical Systems and Critical Infrastructures 5, https://doi.org/10.1007/978-3-031-29301-6_8

PoC	Proof-of-Concept.
PON	Passive Optical Network.
PWM	Pulse Width Modulation.
QoS	Quality-of-Service.
RF	Radio Frequency.
ROS	Robot Operating System.
RU	Radio Unit.
UAV	Unmanned Aerial Vehicle.
UAVs	Unmanned Aerial Vehicles.

Glossary

5G	5th generation mobile network.It enables a new kind of network that is designed to connect virtually everyone and everything together including machines, objects, and devices.
Backhaul	It is a circuit that relays between the core network and the access network. In mobile communications such as smartphones, the line between the core network and a base station equipped with a device called a Base Band Unit that converts IP packets and baseband signals and controls the entire base station.
Network topology	It is a Connection type of computer network. Typical network topologies include bus, star, ring, and full-mesh.
Smart cities	Cities that use digital technologies such as IoT and AI to collect and utilize all kinds of data.
Spatio-temporal fluctuations	The change related to both space and time.
Throughput	A measure of how many units of information a system can process in a given amount of time.
Topic data communications	A mechanism for selectively sending and receiving data based on topic name.The Publisher sends data to a topic with a specific name, and the Subscriber receives data flowing to a specified topic.

1 Introduction

The rapid increase of connected devices including smartphones, M2M (Machine to Machine), and IoT (Internet of Things) devices has resulted in the tremendous growth of global IP traffic. The research and development of the beyond 5G mobile communications system has attracted significant attention to deal with the current and future requirements of mobile applications. The centralized radio access network (C-RAN) architecture has been a popular network architecture for deploying a large number of small cells [1]. In this architecture, the higher layer functions are provided by a central unit (CU) which is installed as a CU pool in a data center. A radio unit (RU) provides radio frequency (RF) components with antenna elements and PHY layer functions to compose a small cell. The use of high frequency bands such as millimeter-wave (mmWave) radio leads to not only the increase in the data rate but the reduction in the cell size. This is because of the attenuation of radio waves increases in accordance with frequency. Since numerous small cells must be densely deployed to meet the traffic demand, it is an important issue to efficiently place mobile base stations [2, 3].

Along with the increase in the amount of the mobile traffic, it has been pointed that the human mobility and lifestyles compose the spatio-temporal patterns of mobile traffic [4]. A typical example of such patterns is the daily commuter movement between residential areas and downtowns during a working day. Such spatio-temporal fluctuations of mobile traffic deteriorate the efficiency of conventional fixed small cells due to the underutilization of cell capacities in off-peak times. To deal with the spatio-temporal fluctuations of the traffic demand, a moving mobile network was proposed [5–7]. The moving mobile network leverages moving small cells to adaptively deal with the fluctuating traffic demand. A moving small cell is a drone, an unmanned aerial vehicle (UAV), or an autonomous vehicle which is equipped with RU functions. This scheme is expected to reduce the number of required nodes by adaptively following the demand distribution to efficiently construct the future mobile networks.

A concept of moving optical network has been one of the emerging research topics [5, 8]. In [5], an adaptive network architecture using moving mobile base stations was proposed to efficiently deal with the fluctuations in mobile traffic demand. In [8], a concept of optically backhauled moving network for local trains was investigated to efficiently provide backhaul links for 5G group mobility where rapidly moving trains carry a large number of user terminals. These concepts have been followed by the investigation for beyond 5G mobile networks such as [6, 9, 10].

In this context, this paper proposes a concept of tethered autonomous moving cells for smart cities. The goal of the tethered moving cells is to provide stable and high data rate communication via optical fiber cables for unstable demands. It assumes applications that require forwarding large amount of data, e.g. high-resolution video, or connecting a large number of nodes that will cause severe interference on wireless networks. Typical examples for such applications are remote control of searching robots in disaster areas and remote environmental monitoring systems using robots. The rest of the paper is organized as follows. Section 2 describes the related works. Section 3 introduces the slackness control mechanism and algorithm to control the tension of fiber cable for tethered nodes. The performance of the proposed scheme is confirmed with computer simulations in Sect. 4. Then, we explain the implementation and experimental results on tethered autonomous cells in Sect. 5. Finally, Sect. 6 provides the conclusions.

2 Related Work

Moving small cells have been a hot research topic for C-RAN to efficiently deploy small cells in wide area. The performance of moving cells was evaluated in [11] considering the scenario where small cells were mounted on public buses. They assumed employing wireless backhaul links for moving cells to communicate with macro cells. Jaziri et al. [12] proposed small cells which were deployed on top of public transportation vehicles such as buses and taxis. Moving small cells which are mounted on vehicles move around crowded streets while carrying traffic generated by the passengers and their vicinities. Vehicle mounted small cells are efficient in crowded areas that compose traffic hotspots inside macro cells. As regards quality-of-service (QoS) in cellular networks with moving small cells, a joint allocation of resource blocks and power was developed in [13]. A vehicular communication system for train and car applications was presented to provide backhaul connectivity for passengers [14]. This system employed a relaying network architecture to support high-data rate using a high carrier frequency while guaranteeing robustness against fast channel variation and unstable handover. However, the major goal of these schemes is to provide connectivity for the passengers in moving enclosed vehicles.

More recently, a tethered flying drone is becoming a promising option for next generation mobile communications. Flying cells are effective to improve throughput of ground users without terrestrial infrastructure, while the battery lifetime and wireless backhaul link capacity were significant limitation. To address this problem, tethered unmanned aerial vehicles (UAVs) has been investigated to provide seamless service over a cable that simultaneously supplies power and data from a ground station at the expense of limited mobility [20]. They presented a comparative performance analysis of tethered UAV and untethered UAV-assisted cellular traffic offloading under heavy traffic conditions. Kishk et al. [18] provided discussion on the potential of tethered UAV for cellular coverage enhancement in rural areas and capacity enhancement in urban areas. Alzidaneen et al. [19] proposed to employ tethered balloons to enhance the end-to-end data rate of ground users with UAVs. The proposed solution was formulated as an integer linear programming problem to optimize the UAV's placement and transmit powers. The optimal deployment of tethered UAVs was studied in [15] to maximize throughput in multi-cell environments. They modeled the problem as a Markov decision process to develop a multi-agent Q-learning algorithm. Kishk et al. [17] formulated the optimization problem for 3D placement of tethered cells considering the maximum length and the launching point of UAVs. Kondo et al. [16] presented the concept of handover for tethered drone cells and a trajectory control algorithm among junction points. They demonstrated the concept using Mavic 2 Pro drones.

However, to the best of our knowledge, tethered ground/floating nodes have not been employed for moving cells. This is mainly because of fiber cables on the ground and water surface. Since an autonomous cell moves around with the connected cable, it is a significant problem to handle the fiber cable. To address this problem, in this paper we propose a slackness control scheme for a moving cell composed with a tethered ground/floating node. The contribution of this paper is to demonstrate a tethered ground/floating cell for the first time.

3 Proposed Tethered Cell

3.1 Concept

The concept of the proposed tethered cells is shown in Fig. 1. The tethered cells provide on-demand connectivity for mobile users. They are connected with junction points via optical fiber cables. Note that the network topology for tethered cells can be either point-to-point or point-to-multipoint such as passive optical network (PON). A junction point is connected to a CU pool with fixed cables. The tethered cables provide high-rate and stable communication links along with long battery lifetime for ground robots and floating nodes.

We also propose the slackness control algorithm to optimally handle the optical cable. The goal of the proposed algorithm is to control the slackness of the optical fiber cables. If the tension exceeds the tensile strength, the cable breaks. Or, excess slackness increases the probability of failures caused by external pressure. The detail of the algorithm is described in the following.

3.2 Fiber Reel

Figure 2 shows the structure of a fiber reel, which is a key idea of the proposed system. It is placed on an autonomously moving node such as a ground robot. A wireless access point is installed on the fiber reel to compose a moving cell. The structure of

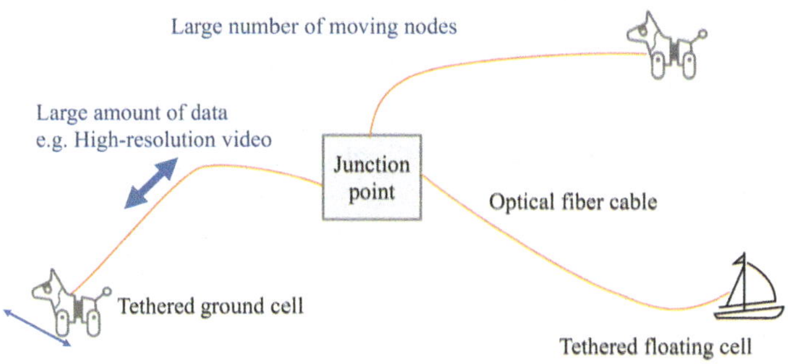

Fig. 1 Concept of tethered cell

Fig. 2 Fiber reel

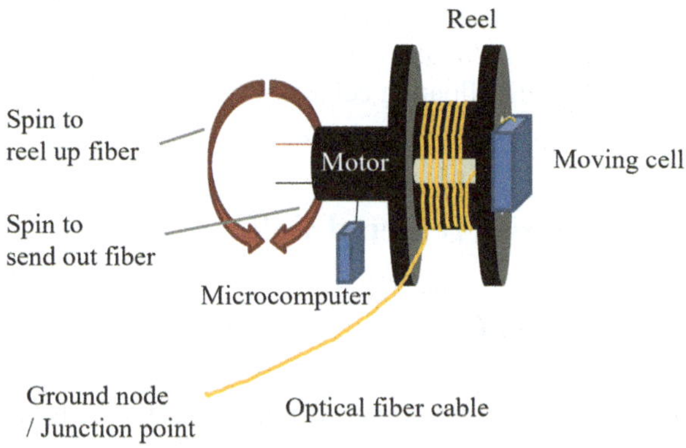

a reel is similar to a fishing reel. A motor is attached to spin to reel up or send out the fiber cable to adjust the cable length. The motor is controlled by a microcomputer.

3.3 Slackness Control Algorithm

3.3.1 Model

The assumed model for formulation is shown in Fig. 3a. Here we assumed a one-dimensional model for simplicity. The position and velocity of a moving node at time t are denoted as $x(t)$ and $v(t)$, respectively. The length of an optical fiber cables between the moving node and the junction point at $x = 0$ is denoted as $L(t)$. The fiber length $L(t)$ can be increased or decreased by rotating the reel according to the movement of the node.

The relationship between $x(t)$ and $L(t)$ is depicted in Fig. 3b. The horizontal tension is described as $T(t)N$. The weight of the fiber is w N/m. The slackness $D(t)$ is approximated with Maclaurin series as

$$D(t) = C(t)\left(cosh\left(\frac{x(t)}{2C(t)}\right) - 1\right) \simeq \frac{wx(t)^2}{8T(t)}, \tag{1}$$

where $C(t) = \frac{T(t)}{w}$. The fiber length $L(t)$ is also approximated as

$$L(t) = 2C(t)sinh\left(\frac{x(t)}{2C(t)}\right) \simeq x(t) + \frac{8D(t)^2}{3x(t)}. \tag{2}$$

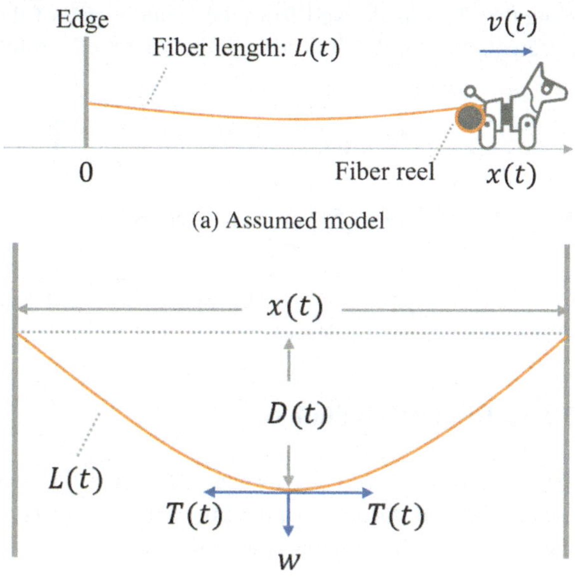

(a) Assumed model

(b) Catenary curve

Fig. 3 Concept of proposed moving optical network

3.3.2 Slackness Control

This section introduces the proposed slackness control scheme. It controls the slackness and the fiber length $L(t)$ by rotating the reel according to the movement of the node.

Let D_{max} and D_{min} denote the maximum and minimum limit of slackness of a cable, respectively. These values are determined considering the safety and stability. D_{min} is generally set as a certain value to reduce the horizontal tension. D_{max} is determined considering the height of the cable to avoid excess slackness because an optical fiber cable is sensitive to external pressure. From (2), the condition for $L(t)$ to satisfy $D_{min} < D < D_{max}$ is

$$x(t) + \frac{8D_{min}^2}{3x(t)} < L(t) < x(t) + \frac{8D_{max}^2}{3x(t)}. \tag{3}$$

This inequality can be written as

$$x(t) + \frac{wx(t)^2}{24T_{max}^2} < L(t) < x(t) + \frac{wx(t)^2}{24T_{min}^2}. \tag{4}$$

where $D_{max} = \frac{wx(t)^2}{8T_{min}}$, $D_{min} = \frac{wx(t)^2}{8T_{max}}$ is defined with (1). T_{max} and T_{min} denote the tension limit for the optical fiber cable; $T_{min} < T(t) < T_{max}$ must always be satisfied.

Note that T_{max} is derived from the catalog spec of the cable. Considering the time differentiation of the fiber length, the fiber reel is rotated to satisfy

$$\left(1+\frac{w^2x(t)^2}{8T_{max}^2}\right)v(t) < \frac{dL(t)}{dt} < \left(1+\frac{w^2x(t)^2}{8T_{min}^2}\right)v(t). \quad (5)$$

With $D_{max} = \frac{wx(t)^2}{8T_{min}}$, (5) can be written as

$$\left(1+\frac{w^2x(t)^2}{8T_{max}^2}\right)v(t) < \frac{dL(t)}{dt} < \left(1+\frac{8D_{max}^2}{x(t)^2}\right)v(t). \quad (6)$$

3.3.3 Distance Limitation

The distance between the moving node and the junction point is limited by the tension and the slackness of the fiber cable. From $D_{max} = \frac{wx(t)^2}{8T_{min}}$ and $D_{min} = \frac{wx(t)^2}{8T_{max}}$, the position of the moving node is described as

$$x(t)^2 = \begin{cases} \frac{8D_{max}T_{min}}{w} & (D_{max} = \frac{wx(t)^2}{8T_{min}}) \\ \frac{8D_{min}T_{max}}{w} & (D_{min} = \frac{wx(t)^2}{8T_{max}}). \end{cases} \quad (7)$$

Given $T_{min} < T(t) < T_{max}$ and $D_{min} < D < D_{max}$, the distance limitation is formulated as

$$x(t)^2 < \frac{8D_{max}T_{max}}{w}. \quad (8)$$

Therefore, multiple junction points and moving nodes should be deployed to deal with the spatio-temporal fluctuations of the traffic demand considering the distance limitation in (8).

4 Numerical Results

4.1 Simulation Condition

The performance of the proposed scheme was evaluated with computer simulations. It was assumed that a connected robot moves in one-dimension as depicted in Fig. 3a. The dynamics of velocity and position of the robot is shown in Fig. 4a. The allowable tension of an optical fiber cable was set as $T_{max} = 60$ N. The weight of an optical fiber cable was set as 6 g/m, and thus $w = 0.006$ N/m. These parameters are typical values for a fiber-optic patch cord. The maximum slackness was set as $D_{max} = 1.0$ m.

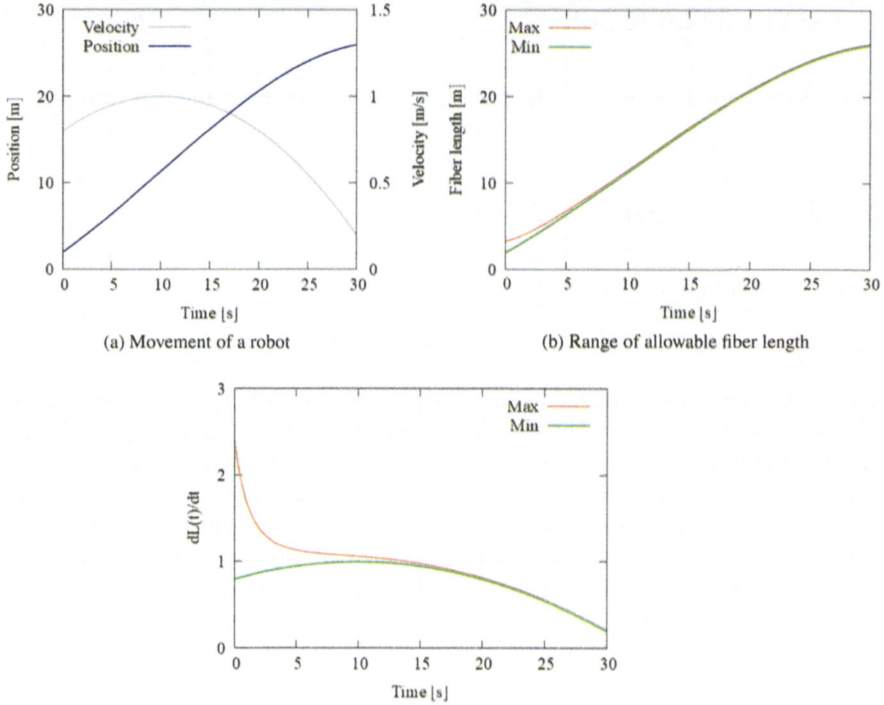

(a) Movement of a robot

(b) Range of allowable fiber length

(c) Range of time differentiation of fiber length

Fig. 4 Simulation condition and results

In this situation, the allowable fiber length and differentiation of it was computed with the proposed scheme.

4.2 Simulation Results

The computed allowable fiber length is shown in Fig. 4b. The fiber length increases in accordance with the movement of the connected robot. The allowable range is wide when the robot is near the edge. This is because it is easy to adjust the excess play in the fiber length in the short-range. Figure 4c shows the range of allowable differentiation of fiber length. It is reasonable to rotate the fiber reel to satisfy this condition. It was confirmed that the slackness of the optical fiber cables for the moving optical network can be controlled based on the computed results.

5 Experimental Results

We confirmed the feasibility of the proposed system with experimental results.

5.1 *Use Case 1: Ground Robot*

5.1.1 Experimental Conditions

Fiber Reel

Here we introduce the implemented PoC (proof-of-concept) system. Figure 5 shows the implemented fiber reel. The end of the optical fiber cable on the robot side was passed through the center of the reel and connected to the router. The router was attached to the side of the reel to prevent the optical fiber cable from breaking. Since the gear ratio of the motor used was 15:1, there was not enough power to turn the reel to match the drive speed. Thus, we used pulse width modulation (PWM) control to match the drive distance and cable length by rotating the reel at high speed and stopping at one-second intervals.

Ground Robot

We employed SoftBank's Cuboid as a ground autonomous robot. Cuboid is a ROS (Robot Operating System) based autonomous driving robot. The fiber reel was installed on the top of Cuboid as shown in Fig. 6. It was connected with the control unit via a serial cable.

We controlled the spinning of the fiber reel based on the movement of Cuboid. ROS employs topic data communications among the components, i.e. they are coupled loosely in the pub/sub model. The Arduino subscribed a certain topic to which the actuator motor of Cuboid published the direction, speed, and distance of movement. When it received subscribing data, it activated a callback function to spin the reel to send out or reel up the fiber cable.

5.1.2 Results

Figure 7 shows the relationship between the fiber length and Cuboid trajectory. The result of the motor power compensation can be seen in Fig. 7a. From the macro perspective, the spinning of the fiber reel was adequately controlled following the trajectory as shown in Fig. 7b.

Figure 8 shows the result of speed test at a certain measurement point. The bottleneck link was the wireless link between the measurement device and the wireless router on Cuboid. Throughput increased as the moving cell approached. This result implies that tethered autonomous moving cell would be effective for mobile users in smart cities.

Fig. 5 Implemented fiber reel

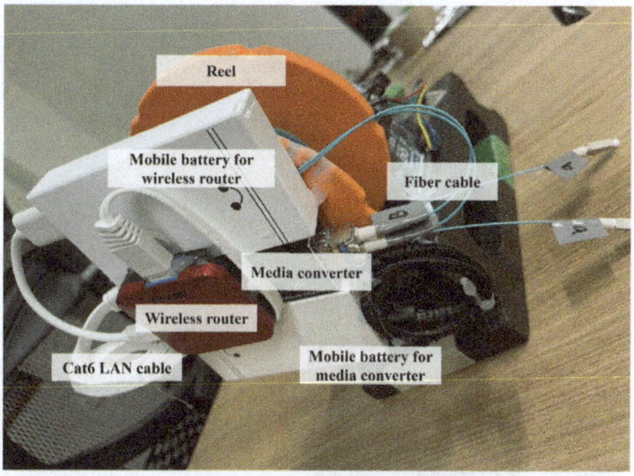

(a) Front

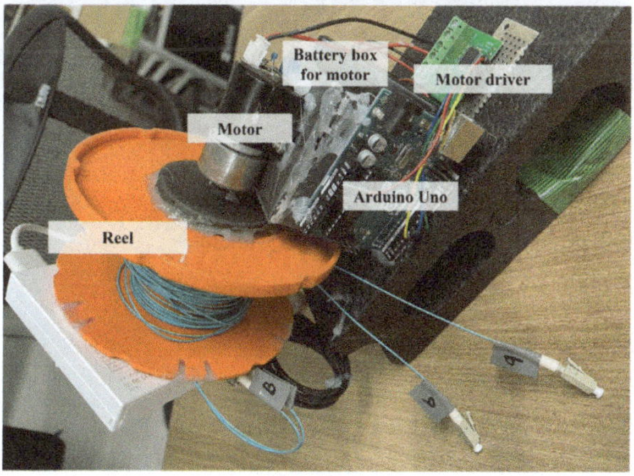

(b) Rear

Although we used a 20-meter fiber cable, the longest distance between the moving cell and the junction point was 15 meters. The difference between the cable length and the movable range was caused by the physical constraints of the actual cable. The optical fiber cable tends to curl easily after being wound up. Thus, it was implied that the actual length of optical fiber cable should be determined based on such physical constraints in addition to the mathematical model.

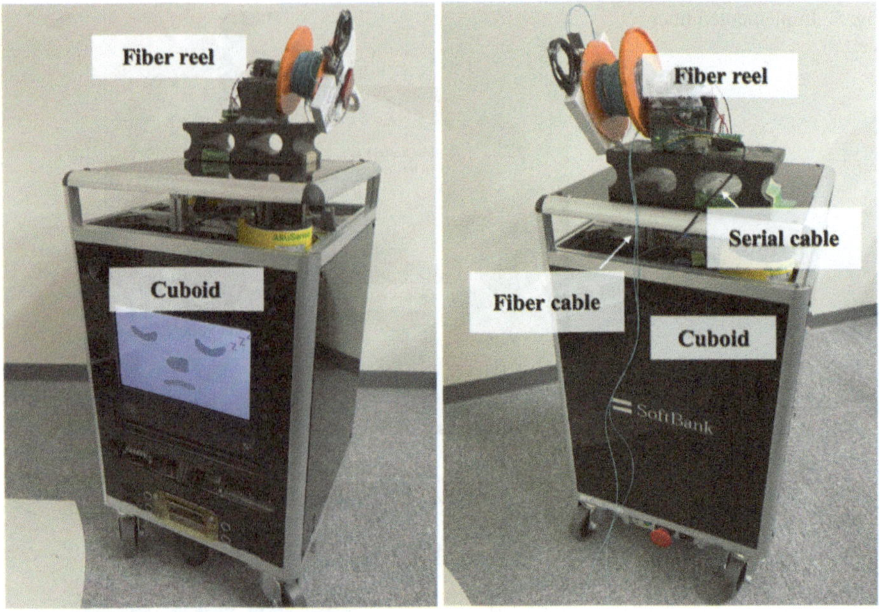

(a) Front (b) Rear

Fig. 6 Autonomous ground cell on Cuboid

5.2 *Use Case 2: Floating Node*

5.2.1 Experimental Conditions

Secondly, we confirmed the feasibility of the moving cell which is floating on water surface. The fiber reel was installed on a floating node as shown in Fig. 9. An underwater drone was mounted to the bottom of the floating node to emulate a movable float. The fiber reel was controlled according to the movement of the drone. We measured the data rate of the wireless link between the wireless router and the handheld device using iPerf.

5.2.2 Results

Figure 10 shows the measured throughput. The throughput increased as the wireless router approached to the handheld device. We confirmed that the installed fiber reel was suitable for adjusting the fiber length on the water surface. Autonomous tethered floating cells will be a strong option for deploying maritime networks which are one of the significant topics for 6G mobile communications.

Fig. 7 Fiber length versus Cuboid trajectory

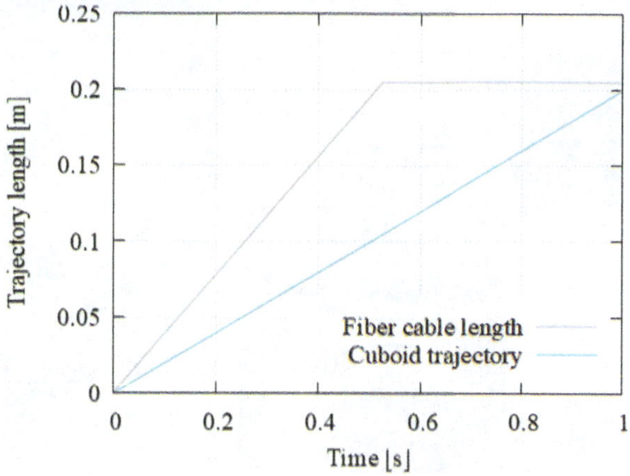

(a) During 1 sec

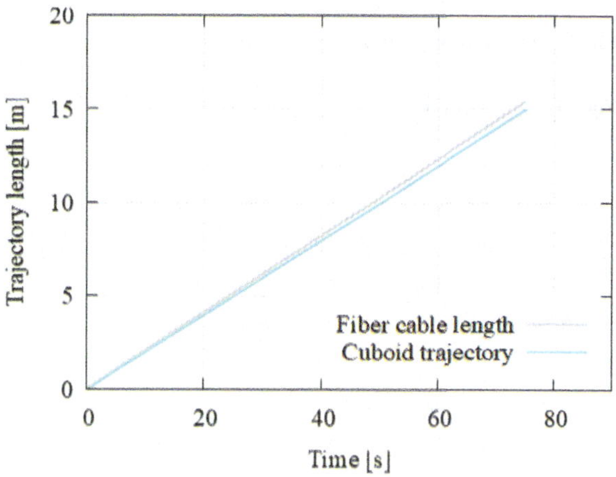

(b) During 80 sec

Fig. 8 Throughput

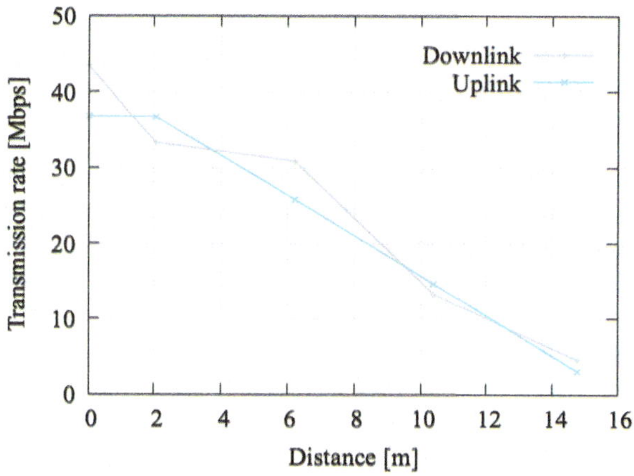

(a) Upper side

(b) Under side

Fig. 9 Autonomous ground cell on underwater drone

Fig. 10 Throughput

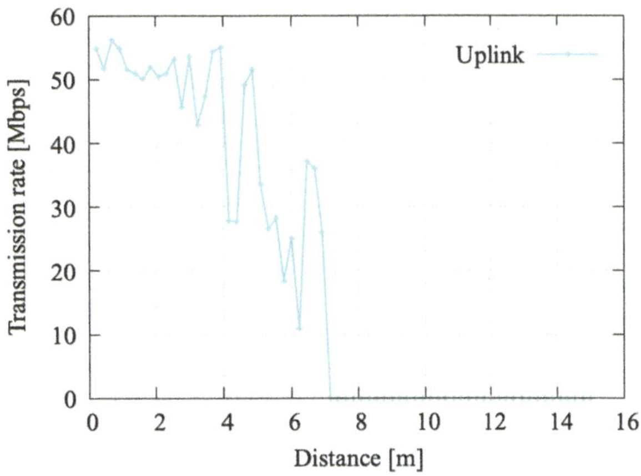

6 Conclusion

This paper proposed a concept of tethered autonomous moving cells for smart cities. The concept of tethered moving networks is a promising solution for various applications including remote control of robots in disaster areas and remote environmental monitoring systems using robots, because of the stable and high data rate communication. This is because a moving tethered cell can dynamically deal with the spatio-temporal fluctuations of the traffic demand to/from mobile devices. Another advantage of a tethered cell is long battery lifetime via wired power transfer. We also proposed the slackness control algorithm to optimally handle the optical cable.

The feasibility of the proposed scheme was demonstrated with the experimental results using a ground robot and a floating node.

Acknowledgements A part of this work was supported by JST, Presto Grant Number JPMJPR2137, and KDDI Foundation, Japan.

References

1. Chanclou P, Anet Neto L, Grzybowski K, Tayq Z, Saliou F, Genay N (2018) Mobile fronthaul architecture and technologies: a RAN equipment assessment. J Opt Commun Netw 10(1):A1–A7
2. Kim J, Chung H, Noh G, Choi S, Kim I, Han Y (2019) Overview of moving network system for 5G vehicular communications. In: 2019 13th European conference on Antennas and propagation (EuCAP), pp. 1–5
3. Mastrosimone A, Panno D (2018) Moving network based on mmWave technology: a promising solution for 5G vehicular users. Wireless Netw 24(7):2409–2426
4. Louail T, Lenormand M, Garcia Cantu Ros O, Picornell M, Herranz R, Frias-Martinez E, Ramasco J, Barthelemy M (2014) From mobile phone data to the spatial structure of cities. Sci Reports 4
5. Nakayama Y, Tsutsumi T, Maruta K, Sezaki K (2017) ABSORB: autonomous base station with optical reflex backhaul to adapt to fluctuating demand. In: IEEE International Conference on Computer Communications (INFOCOM)
6. Nakayama Y, Yasunaga R, Hisano D, Maruta K (2019) Adaptive network architecture with moving nodes towards beyond 5G Era. In: IEEE International conference on communications (ICC)
7. Nakayama Y, Hisano D, Maruta K (2020) Adaptive C-RAN architecture with moving nodes toward beyond the 5G Era. IEEE Netw 34(4):249–255
8. Nakayama Y, Maruta K, Tsutsumi T, Sezaki K (2018) Optically backhauled moving network for local trains: architecture and scheduling. IEEE Access
9. Nakayama Y, Hisano D, Tsutsumi T, Maruta K (2020) Novel c-ran architecture with pon based midhaul and wireless relay fronthaul. In: IEEE consumer communications networking conference (CCNC)
10. Hisano D, Maruta K, Nakayama Y (2020) Low cost c-ran and fronthaul design with wdm-pon and multi-hopping wireless link. In: IEEE consumer communications networking conference (CCNC)
11. Feteiha M, Qutqut M, Hassanein H (2014) Outage probability analysis of mobile small cells over LTE-A networks. In: International wireless communications and mobile computing conference (IWCMC), pp 1045–1050
12. Jaziri A, Nasri R, Chahed T (2016) Offloading traffic hotspots using moving small cells. In: IEEE international conference on communications (ICC), pp 1–6
13. Jangsher S, Li V (2016) Resource allocation in moving small cell network. IEEE Trans Wireless Commun 15(7):4559–4570
14. Noh G, Kim J, Chung H, Kim I (2019) Realizing multi-Gbps vehicular communication: design, implementation, and validation. IEEE Access 19435–19446
15. Lim S, Yu H, Lee H (2022) Optimal tethered-UAV deployment in A2G communication networks: multi-agent q-learning approach. IEEE Int Things J
16. Kondo S, Ota K, Takesita E, Yoshimoto N, Nakayama Y (2022) Autonomous tethered drone cell for IoT connectivity in 6G communications. In: IEEE 95th vehicular technology conference (VTC-Spring) workshop on ICA
17. Kishk M, Bader A, Alouini M (2020) On the 3-D placement of airborne base stations using tethered UAVs. IEEE Trans Commun 68(8):5202–5215

18. Kishk M, Bader A, Alouini M (2020) Aerial base station deployment in 6G cellular networks using tethered drones: the mobility and endurance tradeoff. IEEE Veh Technol Mag 15(4):103–111
19. Alzidaneen A, Alsharoa A, Alouini M (2019) Resource and placement optimization for multiple UAVs using backhaul tethered balloons. IEEE Wireless Commun Lett 9(4):543–547
20. Bushnaq O, Kishk M, Celik A, Alouini M, Al-Naffouri T (2020) Optimal deployment of tethered drones for maximum cellular coverage in user cluster. IEEE Trans Wireless Commun 20(3):2092–2108

Relationship Among Different Types of Input and Model Accuracies in LSTM Driver Models

Kaisei Honda, Shoko Oikawa, and Toshiya Hirose

Abstract This study aimed to investigate the relationship between model inputs and model accuracy using a long short-term memory (LSTM) driver model in a car-following situation. Automated driving is expected to become a fundamental technology in smart cities, and further research and development of driver models that can reproduce the driving operations of individual drivers are expected. The inputs to the model were the ego vehicle velocity, headway distance, relative velocity, log headway distance, time headway (THW), and time to collision (TTC). Six input patterns were set up to investigate the relationship between the inputs to the model and the model accuracies. The experiment was conducted using a driving simulator to simulate traffic congestion on a straight highway and measure the driver's driving data. The results showed that THW was able to construct a more accurate driver model than TTC, and the model accuracy was improved when TTC and THW were used in combination as inputs: the headway distance for TTC and relative velocity for THW. In the future, it will be necessary to construct an automated driving system using a driver model that reflects the characteristics of individual drivers and evaluate its effectiveness.

K. Honda · S. Oikawa
Human Machine System Laboratory, Shibaura Institute of Technology, 3-7-5 Toyosu, Koto-ku, Tokyo 135-8548, Japan
e-mail: md22091@shibaura-it.ac.jp

S. Oikawa
e-mail: oikawa.shouko.g8@sic.shibaura-it.ac.jp

T. Hirose (✉)
Department of Engineering Science and Mechanics, Shibaura Institute of Technology, 3-7-5 Toyosu, Koto-ku, Tokyo 135-8548, Japan
e-mail: hiroset@sic.shibaura-it.ac.jp

R. Shinkuma et al. (eds.), *Advances in Engineering and Information Science Toward Smart City and Beyond*, Engineering Cyber-Physical Systems and Critical Infrastructures 5, https://doi.org/10.1007/978-3-031-29301-6_9

1 Introduction

Intelligent transportation is an essential component of smart cities. In particular, focusing on autonomous mobility, the advantages of automation include an improvement in transportation efficiency, reduction in traffic accidents, and reduction in driving burdens and driver substitution [1]. Japan's Ministry of Land, Infrastructure, Transport, and Tourism (MLIT) has set the goal of Level 4 implementation by 2025, aiming to provide automated driving services in limited areas such as expressways and airports [2, 3]. The World Forum for Harmonization of Vehicle Regulations of the United Nations Economic Commission for Europe (UNECE) established international regulations for automated lane-keeping systems (ALKS) in June 2020, setting safety requirements for Level 3 automated driving systems for passenger vehicles [4]. Automated driving, which has been actively researched and developed as described above, is expected to continue to develop and become a fundamental technology in smart cities. However, focusing on the situation of car-following in a traffic scene, this traffic scene occurs most frequently in urban areas owing to increased traffic volume and large changes in vehicle velocity. These traffic scenes reduce transportation efficiency owing to traffic congestion, increased traffic accidents, and driving stress. It is hoped that automated driving technology will solve these problems in car-following situations. In contrast, when vehicles are controlled automatically, vehicle control may increase driver stress [5]. To solve this problem, the design of automated driving systems should consider the usability of passengers.

It has been reported that usability can be improved if the driving characteristics of human drivers are considered in the control of automated driving [6, 7]. To apply the driving characteristics of drivers to an automated driving system, a method for modeling the ordinary driving behavior of individual drivers can be applied, which can simulate the ordinary driving behavior of individual drivers for control of an automated driving system; thus, the vehicle can be controlled in a way that mimics the ordinary driving behavior of drivers, and consequently, the usability of automated driving is expected to be improved. Thus, driver models that reflect the driving characteristics of individual drivers are considered to contribute to improving the usability of automated driving in smart mobility systems [8], and it is necessary to construct driver models that contribute to the development of automated driving. The long short-term memory (LSTM) neural network proposed by Hochreiter et al. [9] can be used to model longer time-series data by discarding past state information. LSTM is known to improve the vanishing and exploding gradient problems of conventional recurrent neural networks (RNNs), and in recent years, LSTM has been used in a variety of new methods proposed for automated driving systems [10, 11]. Jeong et al. performed vehicle state prediction of surrounding vehicles for motion planning of autonomous vehicles and found that the output results of the LSTM model simulated the human decision-making process [12]. Similarly, LSTM can be used to construct an accurate driver model when the driver's own vehicle is used as the target vehicle. However, LSTM models, which are data-driven methods, need to consider the necessary inputs to simulate driving characteristics, since various types of information affect model accuracy. The main information that influences a driver's judgment during car-following behavior is the relative velocity and distance from the preceding vehicle (PRV) [13]. The relationship between the distance perceived by the driver and the actual distance can be approximated by a logarithmic function, and Fechner's law holds [14, 15]. The time headway (THW) and time to collision (TTC) are representative indices of collision risk judgment for the PRV. In previous studies, these collision risk judgment indices were introduced to construct driver models for car-following [16, 17]. However, few studies have investigated the effects of these input variables on the accuracy of LSTM-based driver models.

The purpose of this study was to investigate the relationship between the inputs to the model and model accuracies using the LSTM driver model in a car-following situation. We constructed an individual driver model using LSTM, which reflects the driving characteristics in a car-following situation (decelerating, following, and accelerating). The inputs to the model were the velocity of the ego vehicle (EGV), the distance between the EGV and PRV, the relative velocity, the log distance between the EGV and PRV, THW, and TTC.

2 Literature Review

Several studies have been conducted to model drivers' car-following behavior. Kehtarnavaz et al. (1998) [18] constructed a car-following model using a back-propagation neural network. Zhou et al. [19] showed that car-following models using RNNs have a better prediction performance because they acquire driver reaction times. RNNs also showed better model accuracy than the classical control model. Naseri et al. [20] investigated psychological perspectives in the structure of car-following models that incorporate TTC. Huang et al. [21] constructed a car-following model based on LSTM-NN and found that it successfully captured asymmetric driving behavior and performed better than other models. Qu et al. [22] constructed a car-following model in CNN-BiLSTM-attention that combines the ideas of LSTM and GM models for trajectory prediction. Li et al. [23] proposed a fusion LSTM-intelligent driver model (IDM) car-following model and a modeling method based on an adaptive Kalman filter. Altche et al. [24] used LSTM to predict vehicle behavior on highways and applied real-world trajectory data from the next generation simulation (NGSIM) dataset for validation. Pengcheng et al. [16] used an autoencoder to extract the main features underlying the time-series data of past driving memories and found that velocity, relative velocity, instant perception time (IPT), and THW were the most relevant parameters, with results showing that they were more important than vehicle distance. Wang [25] investigated the impact of different input settings on model performance in LSTM-based velocity prediction. They found that the past EGV velocity and the velocity of the PRV improved the prediction accuracy. Tong Liu [26] investigated how collision risk judgment indices, such as THW and TTC, affect driver behavior during car-following. Those studies constructed driver models for car-following, but lacked a thorough discussion of the effect of using different types and combinations of inputs on model accuracy.

3 Experimental Methods

This section describes the setup of the experimental participants, experimental apparatus, and LSTM parameters.

3.1 Participants in the Experiment

The participants in this experiment were five male drivers in their 20 s who had a driver's license. In general, data from numerous participants are used in experiments to measure driver characteristics; however, in this study, LSTM was used to construct driver models for each participant in the experiment. As shown in the literature review, LSTM can be used to construct individual driver models. Therefore, the number of participants in the experiment was set to five, and the study focused on the effect of LSTM input information.

We explained the purpose and experimental procedure to the participants and obtained informed consent before conducting the experiment. The study protocol was approved by the ethics committee of the Shibaura Institute of Technology.

3.2 Driving Simulator

In this study, a driving simulator (DS) was used to measure the driver's driving data in a car-following scene (Fig. 1). The screen is cylindrical in shape, with a diameter of 5.0 m, and can present a 360° traffic scene. The roll and pitch motions of the vehicle were simulated by a 6-axis sway device installed in the lower part of the DS. Yaw motion was simulated using a turntable installed in the lower part of the DS. The maximum forward/backward acceleration could be simulated by the DS for a short period of approximately 6.0 m/s^2. The turntable had a movement range of ± 180°, which enables the presentation of yaw motion in the experimental scenario used in this study. The accelerator pedal, brake pedal, and steering wheel simulated the reaction force of the driver's operations, and driving sounds including the sound of other approaching vehicles could be presented through speakers installed inside the vehicle.

Fig. 1 Driving simulator (DS)

(a) Overall picture of the DS

(b) View from driver's seat

3.3 Experimental Scenario

The scenario of the experiment simulated the congestion situation on a straight section of a Japanese highway with two lanes. The driver of the EGV was traveling at 100 km/h on the highway, approached the end of the vehicle queue traveling at 40 to 60 km/h, and decelerated (Fig. 2). The driver drove into the traffic congestion, and after crossing the accident point (Fig. 3), traffic congestion cleared, and the driver accelerated to 100 km/h again. Figure 4 shows the sequence of scenarios. The experiment took approximately 2 min and 30 s per scenario, and the participants drove 10 times in this driving scenario. In all experimental driving conditions, participants were instructed to drive while paying attention to the following four items: (1) not changing lanes and continuing to drive in the travel lane; (2) avoiding collision with the PRV or side walls; (3) obeying the speed limit; and (4) performing ordinary driving with acceleration, deceleration, and control of the timing of pedals or steering wheel.

Fig. 2 Expressway in congested condition

Fig. 3 Accident point that causes congestion

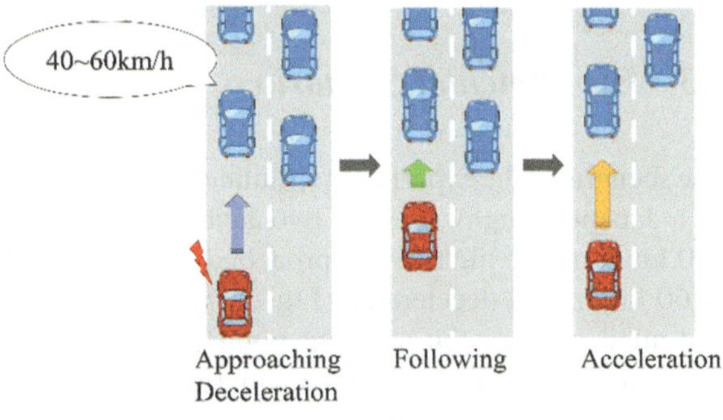

Fig. 4 Scenario sequence (Red car: ego vehicle)

3.4 LSTM

This section describes the modeling of LSTM used in this study to construct the driver model. In this study, driver models were constructed using MATLAB R2021a. In MATLAB, the number of inputs and outputs, number of units, initial learning rate, decay rate of the learning rate, and epochs can be set as hyperparameters. The LSTM consists of an activation function, an input gate, a forget gate, an output gate, and memory cells. The output gate is a simple RNN that feeds back the past outputs to the input. The forget gate outputs the product of the input value and data elements stored in the cell one step earlier. The memory cell and the forget gates play the role of deciding how much of the past information to retain. Similarly, the input gate is responsible for adjusting the retention of current information. This enables LSTM to learn long-term dependencies; therefore, the network can be modeled more accurately because of its deeper layers. The following sections describe the input, output, and LSTM hyperparameter settings.

3.4.1 Input

We selected EGV velocity (V), relative velocity (ΔV), distance between the EGV and PRV (ΔX), logarithmic distance between the EGV and PRV ($log \Delta X$), THW, and TTC as the input variables for the LSTM, which was based on previous studies [16, 17, 25]. The number of input variables to the LSTM has a significant effect on model accuracy. This means that an excessively large or inadequate amount of input information can have a negative impact on the computational and predictive performance of the model [27]. Therefore, in this study, the model was built using the three most frequently used input information sources [22, 23].

We prepared six combinations of input data for training, from Pattern 1 to Pattern 6 (P-1 to P-6), as listed in Table 1. The EGV velocity (V), relative velocity (ΔV), and distance between the EGV and PRV (ΔX) are a combination of inputs used in many car-following models. The logarithmic distance between the EGV and PRV was set based on the assumption that human sensory quantities are perceived in proportion to the logarithm of stimulus quantities [28]. The THW and TTC are typical collision risk judgment indices for PRV. THW is the distance between the EGV and PRV (ΔX) divided by the EGV velocity (V) and represents the relationship between the distance to the PRV at the EGV velocity (V) in terms of time. TTC is the distance between the EGV and PRV (ΔX) divided by the relative velocity (ΔV) and represents the time until collision with the PRV if the EGV is traveling at the current relative velocity (ΔV).

Table 1 shows the patterns of the input variables used in the construction of the driver model in this study. P-1 consists of velocity, relative velocity, and distance

Table 1 Input variables set

Variable pattern	V	ΔV	ΔX	$log \Delta X$	THW	TTC
P-1	○	○	○			
P-2	○	○		○		
P-3	○	○			○	
P-4	○	○				○
P-5	○		○		○	
P-6	○		○			○

between EGV and PRV; P-2 consists of velocity, relative velocity, and logarithmic distance between EGV and PRV; P-3 consists of velocity, relative velocity, and THW; P-4 consists of velocity, relative velocity, and TTC; P-5 consists of velocity, distance between EGV and PRV, and THW; P-6 consists of velocity and distance between EGV and PRV, and TTC. Velocity is used in all patterns because it is considered to be the input information that influences the output, as the following scene in this study involves a large change in driving velocity. Therefore, the velocity was selected as the first input for all patterns. In addition, the distance between the EGV and PRV and the relative velocity were selected as the second input to investigate the influence of the input that indicates the relative relationship with the PRV. As the third input, logarithmic distance ($log\Delta X$), time headway (THW), and TTC were selected as inputs that were used to model car-following behavior.

3.4.2 Output

In this study, we set acceleration as the output variable. In car-following scenarios, the modeled acceleration is a direct representation of the changes in the actions performed by the driver and is used as an output in many studies [29]. In addition, the acceleration rate is related to the control of engine power output, and many classical car-following models set the power output based on the acceleration. The acceleration was also set as the output variable in the optimal velocity model (OVM) [30], IDM [31], and Gazis–Herman–Rothery model (GHR) [32].

3.4.3 Hyperparameters

Because the hyperparameters have a significant impact on the accuracy of the model, in this study, the number of neurons in the middle layer, learning rate, and number of learning epochs were set for modeling. Table 2 lists the hyperparameters used in the study. Five different values were set for each parameter for a total of 125 modeling cases. Driver models can be constructed with high and low model accuracies using the parameters listed in Table 2. These parameters could be used to investigate the model accuracy tendency with a wide range for each input pattern.

The initial values of the weights and biases were fixed for each input pattern such that only the input variables had an impact on the accuracy of the model. The activation function was the ReLU function, the optimizer was Adam. The learning rate decay was set to 0.999, and the learning rate decreased by 0.1% per epoch.

Table 2 LSTM hyperparameters

Parameters	Setting
Hidden units	50, 100, 150, 200, 250
Initial learning rate	0.001, 0.005, 0.01, 0.015, 0.02
Epochs	400, 600, 800, 1000, 1200

3.5 *Model Evaluation Methods and Model Accuracy*

The model was constructed by dividing the data measured with DS into two groups: one for training and the other for evaluation. All data were divided such that 90% of the data were used for training and 10% for evaluation. The driver model was input with data for evaluation, and the acceleration outputs were obtained. The model accuracy was evaluated using the data for evaluation that were not used for training.

3.5.1 Evaluation of Model Accuracy Using Coefficient of Determination

The accuracy of the regression model for the acceleration predicted in this experiment was evaluated using the coefficient of determination (R^2). (1) shows the formula for the coefficient of determination. The coefficient of determination is the ratio of how well the regression model fits the measured value. It is calculated by dividing the residual sum of squares by the total sum of squares, and then subtracting this value from 1. The closer the coefficient of determination is to 1, the higher the model accuracy is. The model accuracy was evaluated by calculating the coefficient of determination using Eq. (1), where the LSTM output is the predicted value and the acceleration measured by the driver's maneuver is the measured value. n was the number of test data points measured for each experimental participant.

$$R^2 = 1 - \frac{\sum_{i=1}^{n}\left(y_i - y_i''\right)^2}{\sum_{i=1}^{n}\left(y_i - y_i\right)^2} \tag{1}$$

(y_i: measured value y_i': predicted value y_i: average of measured values)

3.5.2 Analysis method

For all participants, we calculated the average of the coefficients of determination for each of the six input patterns. We analyzed three items: (1) average coefficient of determination per input pattern; (2) effects of logarithmic distance between EGV and PRV; and (3) effects of THW, TTC, relative velocity, and headway distance between EGV and PRV. For (1), we investigated the overall trends in the input patterns. For (2), P-1 ($V, \Delta V, \Delta X$) and P-2 ($V, \Delta V, log\Delta X$) were compared, focusing on ΔX and $log\Delta X$, using the Mann–Whitney test [33]. For (3), P-3 ($V, \Delta V, THW$), P-4 ($V, \Delta V, TTC$), P-5 ($V, \Delta X, THW$), and P-6 ($V, \Delta X, TTC$) were compared using the Mann–Whitney test. For (2) and (3), we compared the patterns with 625 model accuracies (125 model accuracies × five participants).

4 Results

4.1 Evaluation of Model Accuracy Using Coefficient of Determination

Figure 5 shows an example of the simulation results of the LSTM output and driver-operated acceleration for participant A in the experiment. This shows the results for experimental participant A with 100 units, a learning rate of 0.02, 600 epochs, and P-1 as LSTM input (velocity, relative velocity, and distance between EGV and PRV). The coefficient of determination calculated for the evaluation was 0.9504. The graph in Fig. 5a shows the time on the horizontal axis and the acceleration on the vertical axis, the blue line shows the acceleration in the driver's operation, and the orange line shows the LSTM output. These lines exhibited the same acceleration, and the constructed model was highly accurate. Figure 5b shows the acceleration in the driver's operation on the horizontal axis and the LSTM output on the vertical axis. The model accuracy can be examined according to the magnitude of the acceleration, and if the blue plot is not on the red line, an error occurs. In this case, an error occurred in the driving operation with a large acceleration.

Fig. 5 An example of the prediction results

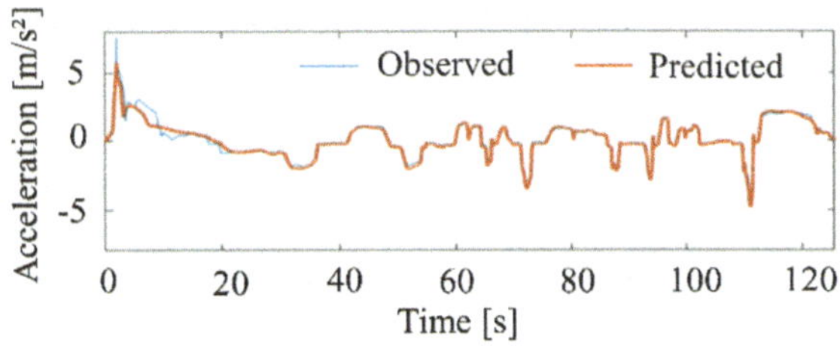

(a) Time series of measured and predicted acceleration variation

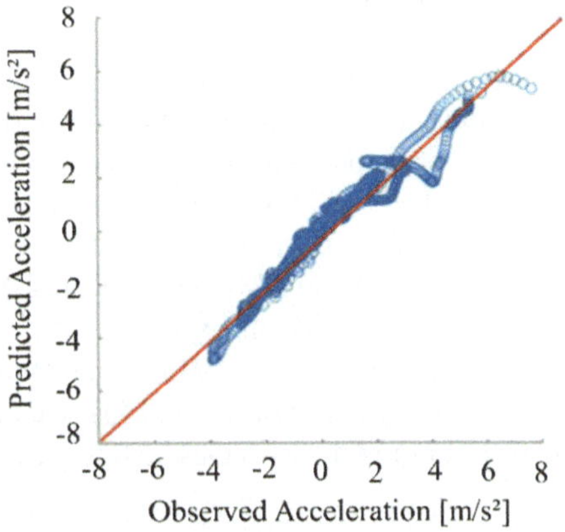

(b) Scatter plot of measured and predicted values

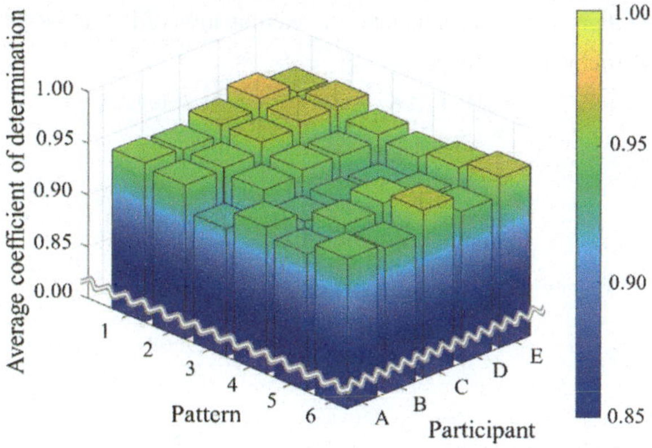

Fig. 6 Average coefficient of determination for each participant

4.2 *Model Accuracy for LSTM Input Patterns*

Figure 6 shows the mean values of the coefficients of determination for the six patterns of the LSTM input for all experimental participants. The three-dimensional graphs showed the average of the results for the 125 parameter combinations.

The average coefficients of determination for each pattern are 0.9604 for P-1 ($V, \Delta V, \Delta X$), 0.9587 for P-2 ($V, \Delta V, log\Delta X$), 0.9452 for P-3 ($V, \Delta V, THW$), 0.9392 for P-4 ($V, \Delta V, TTC$), 0.9464 for P-5 ($V, \Delta X, THW$), and 0.9495 for P-6 ($V, \Delta X, TTC$). P-1 showed the highest average accuracy (0.9604). P-1 and P-2 were higher model accuracy over 0.95. P-4 exhibited the lowest average accuracy (0.9392).

The coefficients of determination for each input pattern tended to be higher than those for the other input patterns, with P-1 ($V, \Delta V, \Delta X$) and P-2 ($V, \Delta V, log\Delta X$) having a coefficient of determination of 0.95 or higher, except for P-2 for participant A. P-1 for participant D had the maximum value of the coefficient of determination of 0.9719 among all patterns and participants. However, P-6 ($V, \Delta X, TTC$) for participants C and E had a higher accuracy than P-1. The coefficients of determination for P-6 for participants C and E were 0.9625 and 0.9647, respectively. The lowest value of the coefficient of determination for each input pattern and participant was P-3 ($V, \Delta V, THW$) for participant A, with a coefficient of determination of 0.9234.

4.3 *Effects of Logarithmic Distance Between EGV and PRV*

Table 3 shows the average coefficients of determination for P-1($V, \Delta V, \Delta X$) and P-2 ($V, \Delta V, log\Delta X$) for each participant in the experiment, and a comparison was made between using ΔX and $log\Delta X$ as the third input. Table 3 shows the average, standard deviation, and difference between the ΔX and $log\Delta X$ for the experimental participants. The maximum value of the coefficient of determination in P-1 was 0.9719 for

Table 3 Average coefficients of determination of P-1 and P-2 for each participant

Participants	Patterns				Difference of ave |(a)–(b)|
	P-1: ΔX		P-2: logΔX		
	Ave. (a)	S.D	Ave. (b)	S.D	
A	0.9537	0.0288	0.9491	0.0350	0.0046
B	0.9518	0.0344	0.9510	0.0326	0.0008
C	0.9616	0.0276	0.9607	0.0250	0.0009
D	0.9719	0.0189	0.9661	0.0222	0.0058
E	0.9630	0.0119	0.9668	0.0218	0.0038

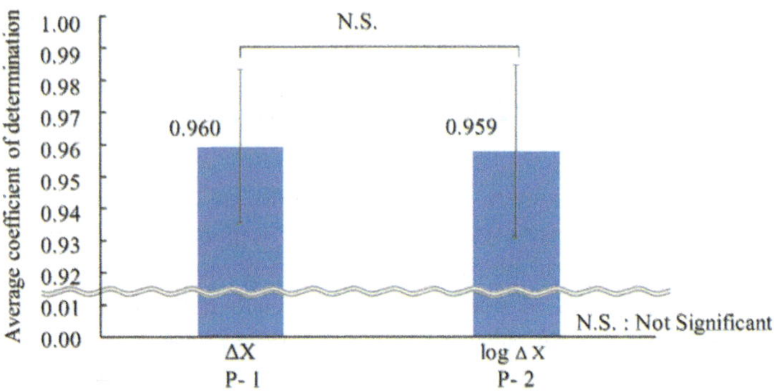

Fig. 7 Comparison of coefficients of determination between P-1 $(V, \Delta V, \Delta X)$ and P-2 $(V, \Delta V, log\Delta X)$ (n = 625 for each pattern)

participant A, and in P-2 was 0.9668 for participant E. The largest difference in the coefficient of determination was 0.0058 for participant D.

Figure 7 shows a comparison of the coefficients of determination for P-1 $(V, \Delta V, \Delta X)$ and P-2 $(V, \Delta V, log\Delta X)$. No significant differences were observed between ΔX and $log\Delta X$.

4.4 Effects of THW, TTC, Relative Velocity, and Headway Distance

Table 4 shows the average and standard deviation of the coefficients of determination for the 125 different patterns for each participant. The coefficients of determination for the participants' accuracy in each pattern ranged from 0.92 to 0.96. Therefore, the Mann–Whitney test was used to calculate the significant differences from P-3 to P-6, and a total of 625 model accuracies were calculated using 125 model accuracies for the five participants.

Table 4 Average coefficient of determination of P-3 to P-6 of the participants

Participants	Patterns							
	P-3: ΔV + THW		P-4: ΔV + TTC		P-5: ΔX + THW		P-6: ΔX + TTC	
	Ave	S.D	Ave	S.D	Ave	S.D	Ave	S.D
A	0.9234	0.0641	0.9417	0.0388	0.9328	0.0412	0.9450	0.0306
B	0.9460	0.0454	0.9328	0.0438	0.9456	0.0331	0.9406	0.0260
C	0.9507	0.0512	0.9360	0.0948	0.9526	0.0431	0.9625	0.0294
D	0.9496	0.0447	0.9346	0.0513	0.9433	0.0554	0.9462	0.0344
E	0.9561	0.0319	0.9510	0.0319	0.9577	0.0203	0.9647	0.0165

The test results showed a significant difference between P-3 $(V, \Delta V, THW)$ and P-4 $(V, \Delta V, TTC)$ ($p < 0.001$), indicating that using THW (P-3) tended to yield a higher accuracy than using TTC (P-4), as shown in Fig. 8. A significant difference was observed between P-4 and P-5 $(V, \Delta X, THW)$ ($p < 0.01$). The combination of headway distance and THW (P-5) tended to provide higher accuracy than the combination of relative velocity and TTC (P-4). A significant difference was observed between the P-3 and P-5 groups ($p < 0.05$). The relative velocity and THW (P-3) showed a higher accuracy than the headway distance and THW(P-5). A significant difference was found between P-4 and P-6 $(V, \Delta X, TTC)$ ($p < 0.001$). Combined with TTC, the accuracy was higher for the headway distance (P-6) than for the relative velocity (P-4).

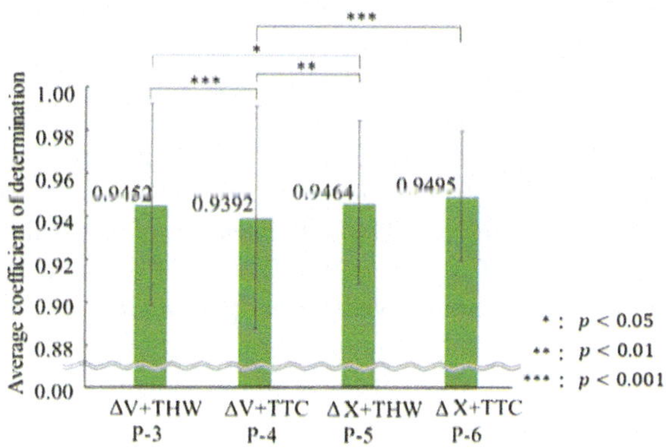

Fig. 8 Comparison of coefficients of determination among P-3, P-4, P-5, and P-6 (n = 625 for each pattern)

5 Discussion

5.1 Model Accuracy When THW and TTC Are Used as Inputs

The test results shown in Fig. 8 indicate that P-3 (V, ΔV, THW) using THW tends to be more accurate than P-4 (V, ΔV, TTC) using TTC. Figure 9 shows the relationship between acceleration and TTC for participant B and the relationship between the measured acceleration and the acceleration obtained from the model output. The coefficients of determination for the acceleration obtained from the model output were 0.940, which is close to the average value, and 0.977, which was the most accurate coefficient for Participant B in the experiment. The mean coefficient of determination for experimental participant B was 0.933, as shown in Table 4. The velocity of the EGV in the car-following scene targeted in this study was often close to the velocity of the PRVs. In other words, when the velocity of the EGV relative to that of the PRV is low, the TTC is extremely long. When the velocity of the PRV is higher than that of the EGV, the TTC is negative; when the velocity of the PRV is lower than that of the EGV, the TTC is positive. As shown in Fig. 9, in the car-following scene, the relative velocity to the PRV changes from positive to negative, so the value of TTC may also alternate between positive and negative values. As shown by the black circles in Fig. 9, the accuracy of the model output tended to decrease when the TTC value switched between positive and negative values. In the car-following scene, the EGV velocity never decreases to zero because the vehicle is in a traveling state, and therefore, the THW value calculated from the velocity is unlikely to take a positive or negative value. Consequently, the tendency observed for TTC was not seen.

In a car-following model, whether TTC or THW should be used as a feature value depends on both the task and model construction method. TTC indicates whether a vehicle is approaching a dangerous distance from other traffic participants and may be used in driver assistance systems such as advanced emergency braking systems [34]. Alternatively, THW indicates the relationship between distance and velocity between an EGV and a PRV. Thus, THW could be useful in the model's task of predicting whether a vehicle is approaching another vehicle. In the car-following model constructed to simulate ordinary driving behavior, THW, which represents the relationship between distance and velocity, was more accurate than TTC. Therefore, it is necessary to study appropriate input information to construct a driver model for car-following under various conditions in the future.

5.2 Effect of Input Features on Model Accuracy

The test results shown in Fig. 8 show that each combination of THW and relative velocity (P-3), TTC, and headway distance (P-6) tended to show high accuracy as input combinations. A previous study investigated the relationship between TTC,

Fig. 9 Relationship between acceleration and TTC for experimental participant B

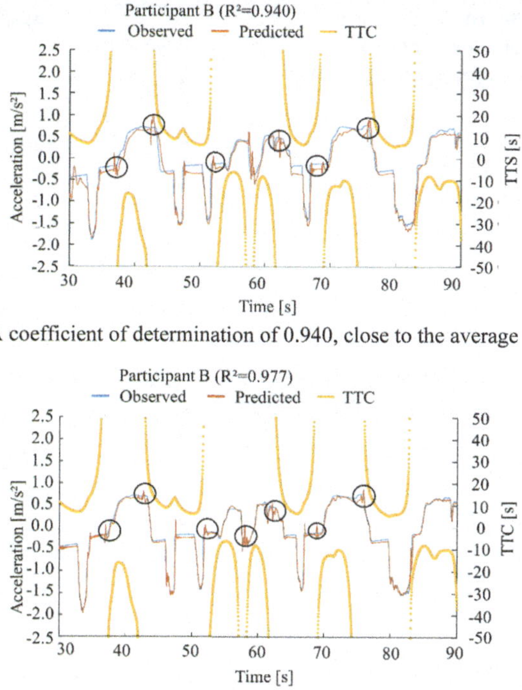

(a) A coefficient of determination of 0.940, close to the average

(b) Highest accuracy with a coefficient of determination of 0.977

velocity, and headway distance [26]. The results showed that TTC tended to decrease as the relative velocity increased, and TTC tended to increase as the EGV velocity increased during the car-following scene. Figure 10 shows the relationship between the headway distance, relative velocity, THW, and TTC for the car-following scene in this experiment. Figure 10a, b show that the headway distance between the EGV and PRV is more similar to THW than to the relative velocity. Thus, the model accuracy of P-3 ($V, \Delta V, THW$) using the relative velocity as an input was higher than that of P-5 ($V, \Delta X, THW$) using the headway distance as an input, and there was a significant difference between P-3 and P-5. In Fig. 10c, d, the relative velocity had a similar effect on TTC as the headway distance. Therefore, as in the THW mentioned above, P-4 ($V, \Delta V, TTC$) used the relative velocity as the input, which was higher than P-6 ($V, \Delta X, TTC$) using the headway distance as the input. This result shows that when different features are used as inputs, the data are compensated, which improves the model accuracy. In addition, when similar features were used as inputs, there was redundancy in the data, which did not improve accuracy.

In general, when superfluous features are set as the inputs, the accuracy of the model is negatively affected [27]. Therefore, to construct a highly accurate driver model, it is necessary to select the appropriate and minimum number of features according to the driving situation [25]. Based on the results of the study, considering

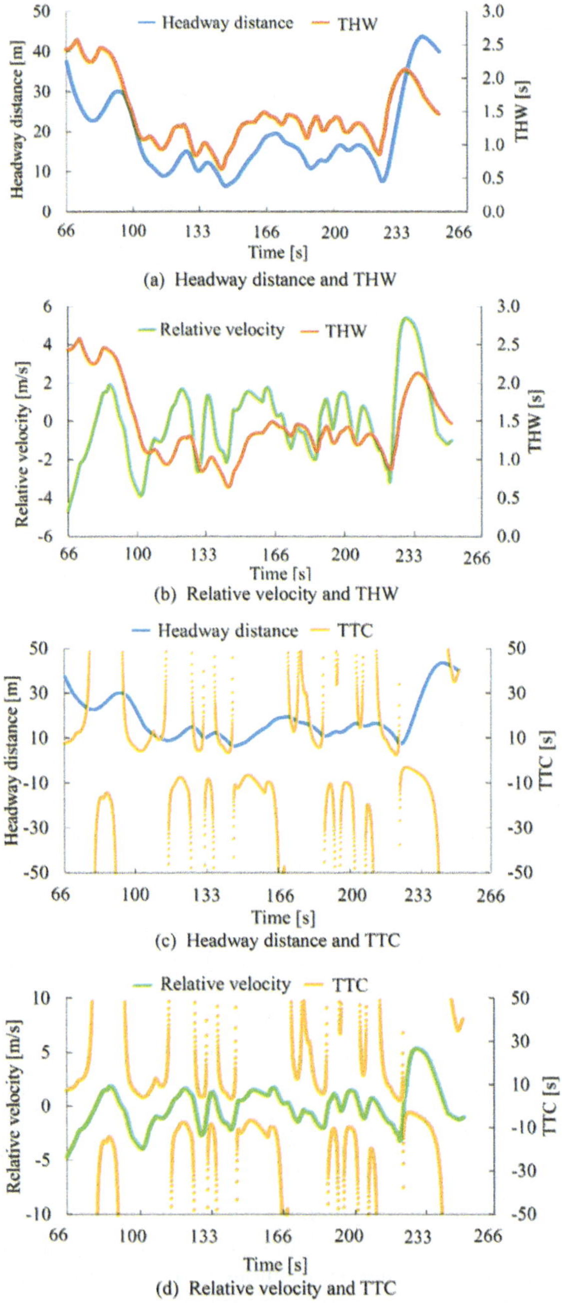

Fig. 10 Time-series variation of headway distance and relative velocity

the combination of input data needed for each scenario may assist in preventing redundancy in the input data.

5.3 Effect of Logarithmic Distance Between EGV and PRV on Model Accuracy

The driver behavior differed significantly between acceleration and deceleration. During acceleration, the driver performs cautious driving maneuvers because the risk of collision increases with increasing vehicle velocity. Therefore, the driver sensitively recognizes the change in the headway distance and performs driving maneuvers. In contrast, during deceleration, the vehicle velocity decreases, the risk of collision decreases by decelerating, and the driver performs careless driving maneuvers. Therefore, changes in the headway distance do not affect driving maneuvers during deceleration [35, 36]. This phenomenon is caused by the difference in reaction speed between acceleration and deceleration maneuvers. In addition, if Fechner's law is simply applied, if the headway distance is long, there is little visual information available about the change in the headway distance, even if the headway distance increases, and if the headway distance is short, there is a large amount of visual information about changes in headway distance. An example of headway distance as visual information is the visual angle of the width of the forward vehicle. In our experimental scenario, the headway distance was distributed between 10 and 400 m. In addition, the distance caused by the driver's driving maneuvers was a short distance of 10 to 20 m. Under the experimental conditions, the logarithmic headway distance did not affect the model accuracy because the headway distance was short during the car-following scenario. However, the logarithmic headway distance input may affect the model accuracy in a traffic scenario in which the vehicle follows the vehicle in front of it with a long headway distance. This may be because, when the headway distance is large, only small changes in visual information occur when the headway distance changes, even if the headway distance increases.

6 Limitations

This study focused on the effects of input information and combinations of inputs on model accuracy. Using feature importance and partial dependence plots, it is possible to quantitatively evaluate the relationship between inputs and outputs. However, feature importance and partial dependence plots are based on the assumption of feature independence. Therefore, it is necessary to examine the appropriateness of applying these methods when relative velocity, headway distance, TTC, and THW are used as inputs to the model.

It is possible that the model accuracy can be improved if the driver's driving characteristics are reflected in the model. In this study, the number of participants in the experiment was limited, and differences in the influence of input information owing to differences in driving characteristics were not investigated. Therefore, it is necessary to investigate the input information for each driver's driving behavior. The consideration of driving characteristics may contribute to improving driver comfort and safety in the vehicle control system of an automated driving system using a personalized driver model in the future.

7 Conclusion

Automated driving is expected to become a fundamental technology in smart cities, and further research and development of driver models that can reproduce the driving operations of individual drivers are expected. The usability of automated driving can be improved if the control of automated driving considers the driving characteristics of human drivers. In this study, we constructed an individual driver model using LSTM, which reflects the driving characteristics in a car-following situation. After constructing the models, we investigated the relationship between the inputs to the model and the model accuracies using the LSTM driver model. The inputs were the velocity, headway distance, relative velocity, logarithmic headway distance, THW, and TTC. The results showed that THW could construct a more accurate driver model than TTC, and the model accuracy was improved when TTC and THW were used in combination as inputs: the headway distance for TTC and relative velocity for THW. In addition, the logarithmic headway distance did not affect the accuracy of the car-following modeling used in this study. In the future, it will be necessary to construct an automated driving system using a driver model that reflects the characteristics of individual drivers and evaluate its effectiveness.

Acknowledgements These research results were obtained partly from commissioned research (No. 06401) by the National Institute of Information and Communications Technology (NICT), Japan.

References

1. Brian P, Michal Č, Yong SZ et al (2016) Survey of motion planning and control techniques for self-driving urban vehicles. IEEE Trans Intel Veh 1(1):33–55
2. SAE (2018) Taxonomy and definitions for terms related to driving automation systems for on-road motor vehicles J3016. SAE International, Warrendale, PA, USA
3. Study Group on Vehicle Safety symposium Ministry of Land, Toward a society without traffic accidents The Future of Vehicle Safety. https://www.mlit.go.jp/policy/shingikai/content/001411236.pdf. Accessed 29 Sept 2022

4. UN. Regulation No. 157 Uniform provisions concerning the approval of vehicles with regard to Automated Lane Keeping Systems. https://unece.org/sites/default/files/2021-03/R157e.pdf. Accessed 29 Sept 2022
5. Franziska H, Patrick R, Matthias B et al (2021) Comfort or Not? Automated driving style and user characteristics causing human discomfort in automated driving. Int J Hum-Comput Interact 37(4):331–339
6. Hammad HA, Ali P, Arne H et al (2022) Impact of perceptual countermeasures on driving behavior at curves using driving simulator. Traffic Injury Prev 20:93–99
7. Marsden G, Mike M, Mark B (2001) Towards an understanding of adaptive cruise control: Towards an understanding of adaptive cruise control. Transp Res Part C Emerg Technol 9(1):33–51
8. Junyu C, Haobin J, Junyan W (2022) Implementation of the human-like lane changing driver model based on Bi-LSTM. Disc Dyn Nat Soc 2
9. Sepp H, Jürgen S (1997) Long short-term memory. Neural Comput 1735–1780
10. Cai J, Jiang H, Wang J (2022) Probabilistic decision-making under Lane Changing Driver Model Based on Bi-LSTM. Disc Dyn Nat Soc
11. Gers FA, Schmidhuber J, Cummins F (2000) Learning to forget: Continual prediction with LSTM. Neural Comput 2451–2471
12. Jeong Y, Kim S, Yi K (2020) Surround vehicle motion prediction using LSTM-RNN for motion planning of autonomous vehicles at multi-lane turn intersections. IEEE Open J Intell Transp Syst 1:2–14
13. Sultan B, Brackstone M, McDonal M (2004) Drivers' use of deceleration and acceleration information in car-following. Transp Res Rec 1883(1):31–39
14. Liang Z, Zhengbing H (2015) A new car following model from the perspective of visual imaging. Int J Modern Phys C 26(8):1550090
15. Takagi T, Taniguchi M, Suzuki S (2012) Weber-Fechner law in road traffic flow. In: Conference on of the society of instrument and control engineers of Japan, pp 1056–1061
16. Pengcheng F, Jingqiu G, Haifeng Z et al (2019) Car-following modeling incorporating driving memory based on autoencoder and long short-term memory neural networks. Sustainability 11(23):6755
17. Rao S, Seitz T, Lanka V et al (2019) Analysis and mathematical modeling of car-following behavior of automated vehicles for safety evaluation. SAE Int J Adv Curr Prac Mobil 876–882
18. Kehtarnavaz N, Groswold N, Miller K et al (1998) A transportable neural-network approach to autonomous vehicle following. IEEE Trans Vehic Technol 42(2):694–702
19. Zhou M, Qu X, Li X (2017) A recurrent neural network based microscopic car following model to predict traffic oscillation. Transp Res C Emerg Technol 84:245–264
20. Naseri H, Nahvi A, Karan FSN (2015) A new psychological methodology for modeling real-time car following maneuvers. Travel Behav Soc 2(2):124–130
21. Xiuling H, Jie S, Jian S (2018) A car-following model considering asymmetric driving behavior based on long short-term memory neural networks. Transp Res C Emerg Technol 95:346–362
22. Dayi Q, Shaojie W, Haomin L et al (2022) A car following model based on trajectory data for connected and automated vehicles to predict trajectory of human-driven. Sustainability 14(12):7045
23. Yufang L, Xiaoding L, Chen R et al (2019) Fusion modeling method of car-following characteristics. IEEE Access 7:162778–162785
24. Florent A, de Arnaud LF (2018) An LSTM network for highway trajectory prediction. In: IEEE international conference on intelligent transportation systems (ITSC) 353–359
25. Jialin W, Shiying D, Qifang L et al (2021) Velocity prediction based on LSTM: impact of different input settings on prediction performance. In: Proceedings of the 2021 5th CAA international conference on vehicular control and intelligence (CVCI)
26. Liu T, Fu R (2018) The relationship between different safety indicators in car-following situations. IEEE Intell Veh Symp (IV):1515–1520
27. Yang SR, Su YJ, Chang YY et al (2019) Short-term traffic prediction for edge computing-enhanced autonomous and connected cars. IEEE Trans Vehic Technol 68(4):3140–3153

28. Keiichi O (2000) Analysis of traffic jam occurrence on a single road of expressway: a model for tracking behavior and future prospects. JSCE Trans 660:39–51. https://www.jstage.jst.go.jp/article/jscej1984/2000/660/2000_660_39/_pdf
29. Mofan Z, Xiaobo Q, Xiaopeng L (2017) A recurrent neural network based microscopic car following model to predict traffic oscillation. Transp Res Part C Emerg Technol 84:245–264
30. Bando M, Hasebe K, Nakayama A et al (1995) Dynamical model of traffic congestion and numerical simulation. Phys Rev E 51(2):1035–1042
31. Treiber M, Hennecke A, Helbing D (2000) Congested traffic states in empirical observations and microscopic simulations. Phys Rev E Stat Phys Plasmas Fluids Relat Interdiscip Top 62(2):1805–1824
32. Herman R, Gazis DC, Potts RB (1959) Car-following theory of steady-state traffic flow. Oper Res 7(4):499–505
33. Jack C (1985) A Wilcoxon-type test for trend. Statist Med 4(1):87–90
34. Meixin Z, Yinhai W, Ziyuan P et al (2020) Safe, efficient, and comfortable velocity control based on reinforcement learning for autonomous driving. Transp Res Part C Emerg Technol 117:102662
35. Edie LC (1965) Discussion of traffic stream measurements and definitions. In: Proceedings of the second international symposium on the theory of road traffic flow, pp 139–154
36. Foote RS (1965) Single Lane traffic flow control. In: Proceedings of the second international symposium on the theory of road traffic flow, pp 84–103

Watch-from-Inside: 3D Sensor System Deployed Inside Building for Road Safety

Hajime Ozawa and Ryoichi Shinkuma

Abstract In the context of smart cities, monitoring systems using three-dimensional (3D) sensors built on light detection and ranging (LIDAR) Light detection and ranging technology have been attracting people's attention as a means of monitoring moving objects such as vehicles and pedestrians at intersections. However, existing works on monitoring for road safety assume that the systems are deployed right near intersections by attaching them to the poles by street or traffic lights, which increases installation and maintenance costs. This paper proposes a watch-from-inside system with 3D sensors deployed inside a building for road safety, WFIWatch from inside for short. When 3D data at an intersection is acquired by a LIDAR sensor through a window between the intersection and the sensor, the window causes deterioration of 3D data caused by attenuation, reflection, and refraction of the laser, which results in degradation of the performance of monitoring moving objects at the intersection. Therefore, we need to take such deterioration into account when we train machine learning models for object detection. The proposed system constructs a conversion model for data obtained without a window. By applying the model to data acquired without a window, we can reflect the deterioration of points caused by the window to the applied data. We construct a conversion model through experiments actually using a LIDAR sensor device and a window. Then, we demonstrate object detection with a machine learning model constructed using the conversion model. We also show the result of object detection for data obtained at an actual intersection through a window.

H. Ozawa (✉) · R. Shinkuma
Shibaura Institute of Technology, Tokyo, Japan
e-mail: ma22033@shibaura-it.ac.jp

R. Shinkuma
e-mail: shinkuma@shibaura-it.ac.jp

R. Shinkuma
Hyper Digital Twins Co., Ltd. (HDT), Tokyo, Japan

R. Shinkuma et al. (eds.), *Advances in Engineering and Information Science Toward Smart City and Beyond*, Engineering Cyber-Physical Systems and Critical Infrastructures 5, https://doi.org/10.1007/978-3-031-29301-6_10

Acronyms

LIDAR	Light detection and ranging
ML	Machine Learning
WFI	Watch from inside

Glossary

Conversion model The conversion model is a conversion model for 3D data. It is applied to the 3D data to convert it into data suitable for constructing an ML model.

Deep learning Deep learning is a type of machine learning that works with algorithms based on the structure and function of the brain, called artificial neural networks.

KITTI dataset KITTI dataset is one of the most popular datasets for use in mobile robotics and autonomous driving.

LIDAR LIDAR is an acronym of "light detection and ranging". It is a method for determining ranges by targeting an object or a surface with a laser and measuring the time for the reflected light to return to the receiver.

Object detection Object detection is a computer vision technique for locating instances of objects.

Point cloud A point cloud is a set of discrete sets of data points in space. A point cloud measures many points on the external surface of a surrounding object.

Smart cities A smart city is a new city that uses advanced Internet of Things (IoT) technology to improve operational efficiency, share information with the public and provide a better quality of government service and citizen welfare.

Smart monitoring Smart monitoring is a key component of a smart city, collecting information for efficient management of infrastructure and resources.

Spline completion Spline completion is a method of interpolating data using polynomials. It is a method that divides a section of data at a point and interpolates using a different polynomial for each section.

1 Introduction

The global population growth and urbanization have been creating new social and economic challenges. A smart city is a platform which aims to provide various various public services for citizens by using various information and communication technologies [1]. Smart monitoring systems for smart cities will improve our living conditions in terms of comfort and safety [2]. For example, issues related to transportation include the increasing number of cars, motorcycles, bicycles, and other road users, which increases the risk of accidents and traffic congestion [3, 4]. In smart cities, moving objects on the road are monitored in the hope of helping the prevention traffic accidents. In particular, two- and three-dimensional (2D and 3D) image-sensor technologies such as cameras and light detection and ranging (LIDAR) can be used for such monitoring [5–8]. 3D sensors enable 3D object detection in the daytime and nighttime without being affected by light. They are also beneficial from a privacy aspect: they do not recognize personal information such as people's faces and license plates. We note that this paper assumes only point-cloud format data acquired by LIDAR sensors as 3D data.

One popular line of research for 3D sensor monitoring focuses on deploying a 3D sensor or multiple 3D sensors outside at an intersection to detect and track moving objects such as pedestrians and vehicles [9]. On the basis of 3D data acquired by these 3D sensors, pedestrians and vehicles are detected and tracked using background filtering, object clustering, pedestrian and vehicle classification, and object tracking [10]. Z. Zhang et al. discussed a way of extracting a background on the basis of the relevance of point data in 3D data for vehicle detection [11]. These prior pieces of work assume that a 3D sensor system is deployed right near intersections by attaching them to the pole of a street light or traffic light, which increases installation and maintenance costs. These system costs need to be reduced to accelerate the deployment of this kind of system extensively over an urban area.

This paper proposes a watch-from-inside system with 3D sensors deployed inside a building for road safety, WFIWatch from inside for short. The proposed system does not need 3D sensors to be attached to outside poles, thus reducing installation and maintenance costs. A possible drawback of the system could be that there is a window between the 3D sensors and the objects and roads observed by the sensors. Therefore, the 3D data acquired from the sensor through the window may cause data deterioration such as attenuation, reflection, and refraction. The proposed system constructs a conversion model for the data acquired without the window. The model is applied to data acquired without a window and reflects the through-window deterioration to the applied data. Such data modified by the conversion model can be used for training data for an MLMachine Learning model for object detection through a window. Using a real LIDAR sensor device, we construct a conversion model for the proposed system and apply it to an ML model for object detection. Light detection and ranging We then present the detection performance of the ML model for target objects placed outside of a window. We also show the result of object detection for data obtained at an actual intersection through a window.

This paper is organized as follows. Section 2 reviews related work on object detection from point cloud. Section 3 presents the system model and the methodology of the proposed system. Section 4.1 presents experiments for constructing a conversion model for the effect of windows and an ML model for object detection using the model. Section 4.4 then presents the evaluation of object detection using the ML model. Section 5 then presents the result of object detection for data obtained at an actual intersection through a window. Section 6 concludes the paper.

2 Related Works

Object detection has been well studied in the field of computer vision. Recently, 3D object detection from point cloud data has attracted researchers' and engineers' attentions as LIDAR sensors become popular [12]. Light detection and ranging By the architecture of the detection process, 3D object-detection methods can be categorized into two: region-proposal based and single shot [13].

Region-proposed methods first produce proposals as possible regions containing objects. They then extract region wise features to determine the category label of each proposal. According to their approaches, region-proposed methods can further be divided into three categories: multi-view based, segmentation-based, and frustum-based [12, 13]. Multi-view based methods fuse proposal-wise features from different view maps, e.g., LIDAR front view, Bird's Eye View (BEV), and image, to obtain 3D rotated boxes. Segmentation-based methods first leverage existing semantic segmentation techniques to remove most background points, and then generate a large amount of high-quality proposals on foreground points. As a segmentation-based method, Shi et al. proposed PointRCNN which performs a point-based process using PointNet++ in all stages to avoid information loss due to quantization [14]. Frustum-based methods first leverage existing 2D object detectors to generate 2D candidate regions of objects and then extract a 3D frustum proposal for each 2D candidate region. Other region-proposal methods than multi-view, segmentation, and frustum based methods have been also investigated. Shi et al. proposed Part-A2, a neural network that consists of two stages: an object part-aware stage and an object-part aggregation stage [15]. The part-aware stage applies sparse convolution and sparse deconvolution to the input point cloud to predict intra-object part locations and generates 3D proposals. The part-aggregation stage conducts RoI-aware point cloud pooling to aggregate the object part information for each 3D proposal. The object part aggregation network is applied to score boxes and it refines the locations of objects on the basis of the features of the parts and information from the part-aware stage. Shi et al. also proposed PointVoxel-RCNN (PV-RCNN) which performs a voxel-based process similar to Part-A2 net in the first stage and then a point-based process using PointNet-based networks in the second stage [16].

Differently from region-proposed methods, single shot methods directly predict class probabilities and regress 3D bounding boxes of objects using a single-stage network. They do not need region-proposal generation and post-processing; they are faster than region-proposal methods. Single shot methods can be further categorized into three categories according to the type of input data: BEV-based, discretization-based, and point-based [13]. BEV-based methods mainly take BEV representation as their input. Discretization-based methods convert a point cloud into a regular discrete representation, and then apply CNN to predict both categories and 3D boxes of objects. For example, VoxelNet represents point clouds using voxel representation and applies it to 3D CNN [17]. PointPillars represents a point cloud using a pillar representation and applies it to a 2D CNN [18]. Point-based methods directly take raw point clouds as their inputs. As example discretization-based types are SECOND and PointPillars, which convert a point cloud into a regular discrete representation. In particular, SECOND takes raw points as input, divides the point cloud into equally spaced voxels using VoxelNet, and encodes the features in each voxel into a 4D tensor [19]. PointNet is used to learn the features of the point cloud organized into pillars, and the learned features are encoded as pseudo-images. It uses PointNet to learn the features of the point cloud that has been organized into pillars and encodes the learned features as a pseudo image. In PointNet, a 2D object-detection pipeline is then used to determine 3D bounding boxes.

There are two methods of generating proposals and bounding boxes: anchor-base and anchor-free [15], The anchor-base method defines anchors, i.e., the average box size for each object, in advance and applies them sequentially to the feature map. The anchor-free method generates bounding boxes directly from points belonging to the target objects by regressing the difference of each point from the center. Point-RCNN uses the anchor-free method, while SECOND, Pointpillars, and PV-RCNN use the anchor-base method. Part-A2 uses both methods.

3 Proposed System

3.1 System Model

Figure 1 shows a system model of the proposed system. It consists of 3D sensors (called sensors hereafter), an edge device (called device hereafter), a storage, an edge computer, a monitor, and a cloud computer. The edge device consists of a receiver and a recorder that receive and record data from sensors. The edge computer consists of an extractor, detector, and visualizer, which performs detection and visualization using machine learning from the extraction of stored data.

In the system, the sensors acquire 3D data and send them to the device. The receiver of the device receives the 3D data from the sensors. The recorder of the device records the 3D data and stores it in the storage. The extractor of the edge computer retrieves an MLMachine Learning model from the cloud computer for object detection from point cloud data. The ML model is obtained by applying the conversion model to the dataset and using the applied dataset for training. The detector of the edge computer performs object detection for moving objects such as cars, bicycles, and pedestrians even through windows. The visualizer of the edge computer provides a visualized result of object detection to the monitor. It is expected that the proposed system performs object detection accurately by using an ML model produced with a conversion model that fits the condition of data acquisition.

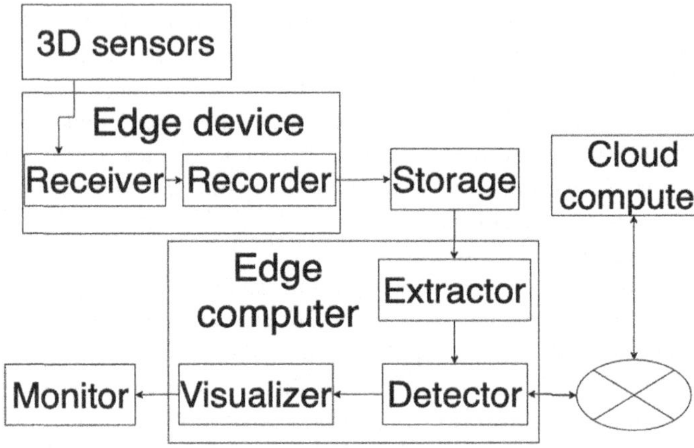

Fig. 1 System model

3.2 *Methodology*

The proposed system is beneficial in terms of installation and maintenance costs compared with the outdoor system because (1) it does not require the outdoor system to be downsized; the window side has more space for deployment than outdoor poles, and (2) the system can easily obtain power from the building. However, a possible drawback could be the deterioration in 3D data caused by the window, due to the degraded accuracy of object detection using 3D data and the ML model.

To overcome this drawback, we suggest the methodology illustrated by the flowchart in Fig. 2. First, the 3D sensor acquires data through and without window glass. Then, these data are compared to measure the degradation caused by the glass. Next, a conversion model is constructed on the basis of the measured data. Then, converted data is obtained by applying the constructed conversion model to training datasets. Then, training is performed using the converted data to construct a MLMachine Learning model for object detection. Finally, the ML model is used to evaluate the accuracy of object detection for the test dataset. A general form of the conversion model is given as:

$$p' = w(p).$$

p' is the 3D data after conversion, $w()$ is the conversion model, and p is the 3D data before conversion. It is difficult to represent $w()$ mathematically; $w()$ could be a non-linear and complicated function because the number and the structure of points changes drastically between p and p'. In particular, change of data structure is problematic in machine learning. The proposed system handles the problem with data structure by converting point-cloud data to 3D voxels. By using the form of

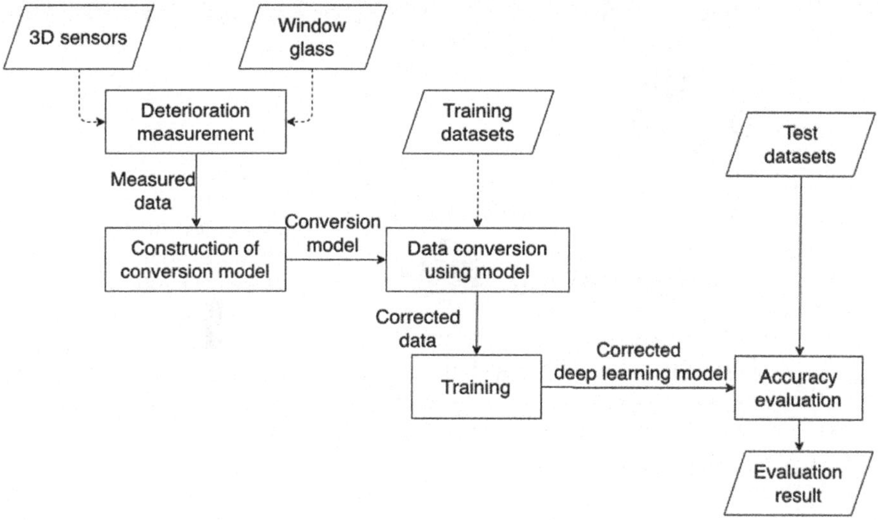

Fig. 2 Flowchart of methodology

3D voxels, we can produce an attenuation map that represent how many points are reduced for each voxel after conversion as a simple form of $w()$. We note that in Sect. 4, since we can ignore the height, point-cloud data is convoluted to 2D pixels.

4 Evaluation

4.1 Experiment for Data Collection

Through experiments, we obtained three types of data: data through a window for conversion, data without a window for conversion, and data through a window for evaluation.

4.1.1 Experiment System

Figure 3 shows the system for the experiment. It consists of only a sensor, a device, storage, and a tripod. The sensor acquires point cloud data as 3D data and sends them to the device. The edge device stores the received data in storage. The sensor is mounted on a tripod. We used a Velodyne VLP-16 [20] and Jetson Nano [21] as the sensor and the device, the specifications of which are summarized in Tables 1 and 2, respectively. Figure 4 shows the experiment site. The location of the experiment was

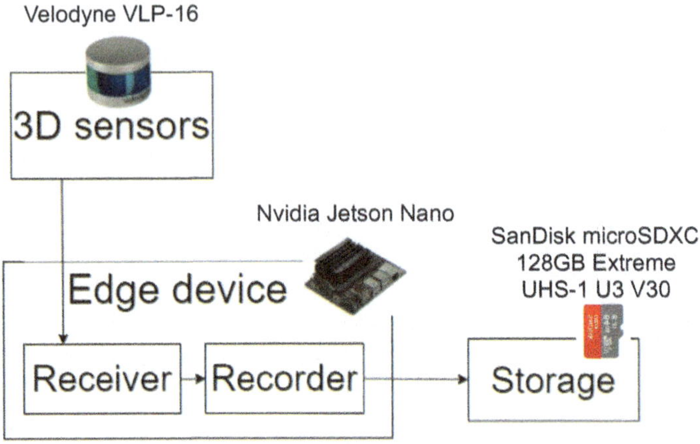

Fig. 3 Experiment system

Table 1 Specifications of velodyne VLP-16

Description	Value
FOV vertical	−15°∼ +15°
FOV horizontal	360°
Measurement distance	0.5∼100 m
Number of obtainable points	Approximately 300,000 per second
Frame rate	10 fps

Table 2 Specifications of jetson nano

Description	Value
CPU	Quad-core ARM A57 @ 1.43 GHz
Memory	4GB 64-bit LPDDR4 @ 25.6 GB/s
Storage	16GB eMMC 5.1
OS	Ubuntu 18.04

the flower garden and the public space in front of it at Toyosu campus, Shibaura Institute of Technology, Koto-ku, Tokyo, Japan.

4.1.2 Data Through Glass for Conversion Model

We first obtained data through glass for the conversion model using the experiment system described in Sect. 4.1.1. Figure 5a shows the placements of the sensor and the object. The thickness of the glass we used was 2 mm. The sensor was placed at 0.5 m from the glass. The sensor was fixed horizontally 1.9 m from the ground. The sensor

Fig. 4 Experimental site

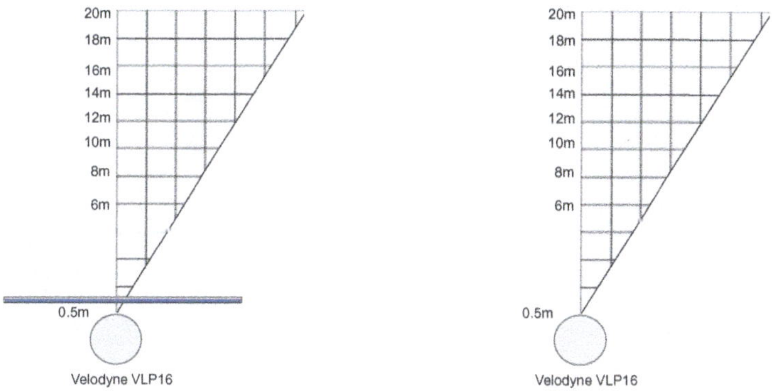

(a) Through glass

(b) Without window

Fig. 5 Sensor and object placement

was placed at a 15-degree angle to the ground. We used a cardboard box, the size of which was 1030 × 230 × 680 mm, as the object for data acquisition. The object was placed at one of the grid-point in the figure. After finishing the data acquisition for the point, we then moved the object to another grid-point. The points were aligned every 2 m in the area shown in the figure, the depth of which was 6–20 m from the sensor and the angle of which was 30° for both the right side.

4.1.3 Data Without Glass for Conversion Model

We then obtained data with no glass for the conversion model using the experiment system described in Sect. 4.1.1. Figure 5b shows the placements of the sensor and the object. Basically, the grid-point were aligned the same as in Sect. 4.1.2, though there was no glass between the sensor and the points. We used the same cardboard box as in Sect. 4.1.2.

4.1.4 Data Through Window for Evaluation

The last data we obtained through the experiment was used for the evaluation shown later in Sect. 4.4. Figure 6 shows the placement of the sensor and the object. Note that, differently from the experiments described in Sects. 4.1.2 and 4.1.3, the object in this experiment was a pedestrian who did not move. The points of the grid were

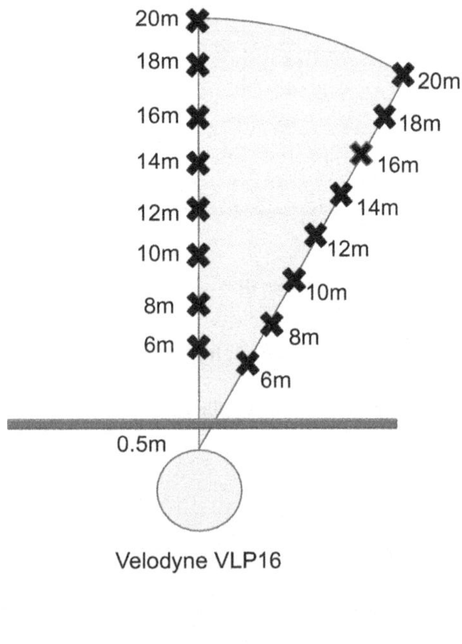

Fig. 6 Sensor and pedestrian placement for data for evaluation

aligned every 2 m from 6 to 20 m in the depth direction from the glass. The degrees were 0 and 15 for the left side. The thickness of the glass was 2 mm.

4.2 Construction of Conversion Model

To construct the conversion model, we used the two types of data we obtained from the experiments described in Sects. 4.1.2 and 4.1.3. As seen in Fig. 5a and b, the two types of data contained point cloud data for the object at each point of the grid, which was common between the two experiments. Namely, by calibrating the coordinates of the two types of data with each other, the number of points obtained for each point could be compared. As we mentioned in Sect. 1, degradation of data acquired through a window could be caused by attenuation, reflection, and refraction of the laser. In this study, we focus only on attenuation, which we confirmed has obviously a large influence on the data acquired through a window. We define the number of points obtained through the window divided by those obtained without the window as the "attenuation" ratio. Figure 7 shows the attenuation ratio at each point of the grid. We used 300 frames for each of the two types of data. For the 0-degree direction, the attenuation ratio basically decreased as the distance increased. However, for other degrees, the attenuation ratio did not always depend only on the distance, which may be caused by the refraction of the LIDAR signal through the glass. Light detection and ranging

As shown in Fig. 7, the attenuation ratio is available only for a discrete grid-point. Basically, the attenuation tends to increase as the distance or the angle increase. However, the tendency is not simple; since the laser unit of the LIDAR sensor rotates, the effect of the window would not be linear for the distance and asymmetric for the angle. Therefore, we applied 2D spline completion to the result in this figure to make it spatially continuous. The spline completion does not apply a single high-degree polynomial to all values at once, it applies a lower-degree polynomial to each value [22]. The spline completion can avoid the problem that a single high-degree polynomial interpolation for equally placed values like those obtained in this experiment would cause oscillations at the edge of the interval. In this work, we do not directly show how accurately the spline completion works because the proposed system does not limit the conversion model into spline completion. Instead, we will show the result of object detection, which suggests how the spline completion works as the conversion model. We used SciPy for the 2D spline completion [23]. Figure 8 shows the attenuation ratio for each location after the 2D spline completion. From this continuous model, the attenuation ratio for each location is available with 1 mm of granularity.

Fig. 7 Attenuation rate for each installation coordinate

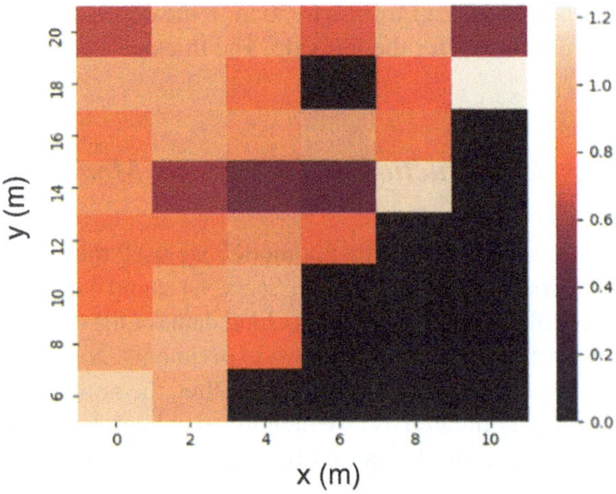

Fig. 8 Map with corrected attenuation ratio

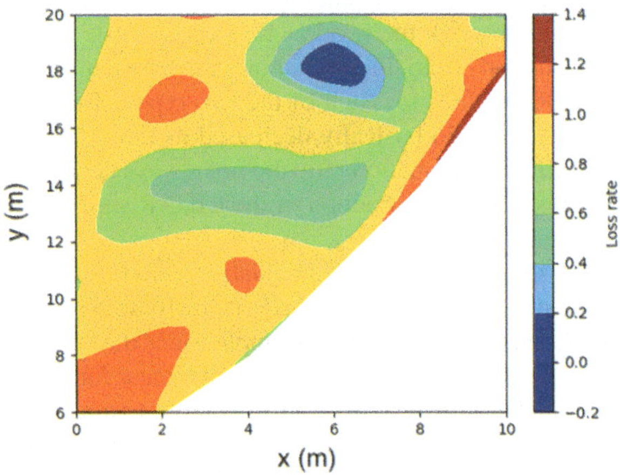

4.3 Construction of ML Model

To construct MLMachine Learning models, training data is needed. In particular, deep learning models, which are used for object detection from image data including 3D point cloud data acquired by LIDAR, need a large amount of data for training. Light detection and ranging KITTI is a well-used dataset for object detection for research on autonomous-driving technology [24]. We used this dataset as training for three reasons: (1) Velodyne's LIDAR was used as we use in our experiment, (2) it contains data for detecting pedestrians, cars, and bicycles, and (3) it has been widely used for research on mobility applications including road safety. The original data of the KITTI dataset was obtained from the onboard sensor of a vehicle; the data were not obtained through glass. We applied the conversion model described in Sect. 4.2 to it. We set the traveling direction of the vehicle of the KITTI dataset

Table 3 Specifications of workstation used as cloud computer

Physical machine	
CPU	Intel Xeon Silver 4210 2.20 GHz 10 Core 20 Thread
GPU	NVIDIA Quadro RTX A6000
Storage	480-GB
Memory	64-GB
OS	Ubuntu 20.04
Docker container	
OS	Ubuntu18.04
Library	CUDA11.4
	Nvidia-driver-460.91.03
	Python 3.8.10
	libpcl-1.8.1

to 0° corresponding to the spatial model of the attenuation ratio shown in Fig. 8. We produced a new training dataset by reducing the number of points for each location in the subset of the KITTI dataset using the attenuation ratio.

Using the training dataset explained above, we constructed a MLMachine Learning model for object detection. We used a deep learning library from point clouds called OpenPCDet [25]. We used different types of models: SECOND, Part-A2-Free, Part-A2-Anchor and PointRCNN; SECOND, Part-A2-Free and Part-A2-Anchor a voxel-based model, while PointRCNN is a point-based. We list the specifications of the server used for the training in Table 3.

4.4 Evaluation Result

4.4.1 Visual Examples of Object Detection

This section presents the evaluation result for object detection using the evaluation data obtained from the experiment described in Sect. 4.1.4.

We first present a visual example of object detection for SECOND in Fig. 9. As seen in Fig. 9a, without our conversion model, multiple pedestrians and even vehicles were detected though in fact, only a pedestrian existed. As seen in Fig. 9b, a pedestrian was successfully detected when our conversion model was used.

We next present a visual example of object detection for Part-A2-Anchor in Fig. 10. As seen in Fig. 10a, without our conversion model, no object was detected. As seen in Fig. 10b, a pedestrian was successfully detected when our conversion model was used.

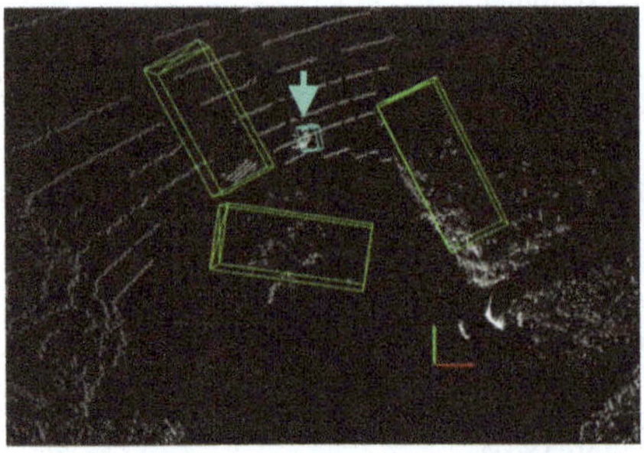

(a) Without our conversion model

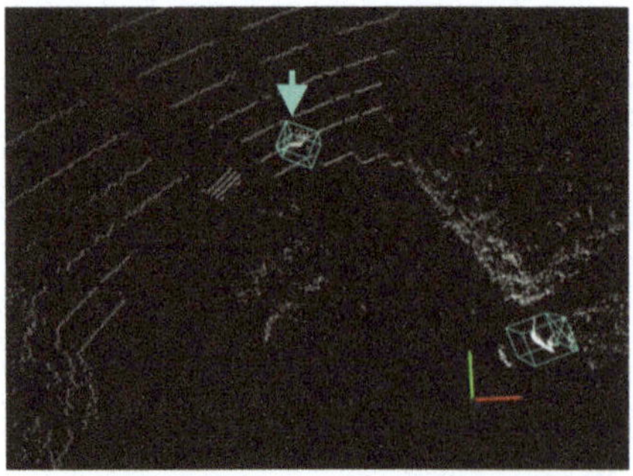

(b) With our conversion model

Fig. 9 Visual example of detection using SECOND for evaluation data. The green bounding box indicates the detection result of a car, and the light blue bounding box indicates the detection result of a pedestrian. Distance of object was 6 m from sensor. Arrows indicate ground truth of pedestrian's position

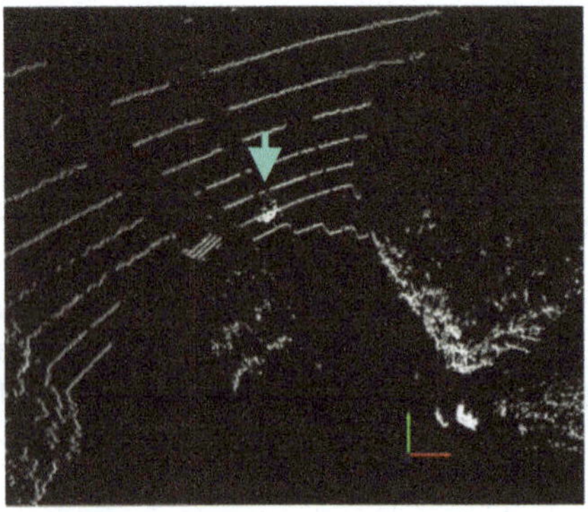

(a) Without our conversion model

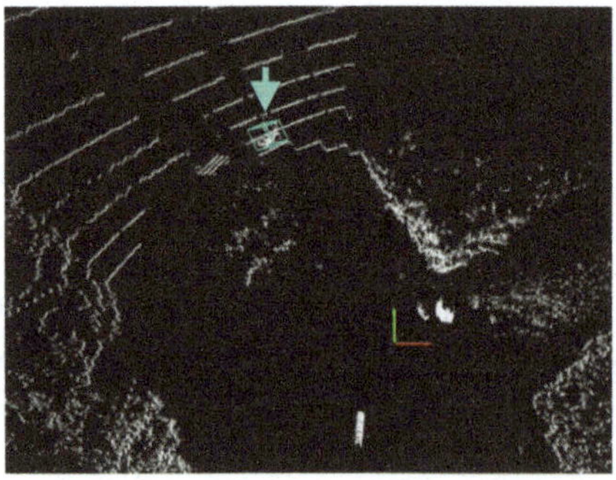

(b) With our conversion model

Fig. 10 Visual example of detection using Part-A2-Anchor for evaluation data. The green bounding box indicates the detection result of a car, and the light blue bounding box indicates the detection result of a pedestrian. Distance of object was 6 m from sensor. Arrows indicate ground truth of pedestrian's position

We next present a visual example of object detection for Part-A2-free in Fig. 11. As seen in Fig. 11a, without our conversion model, no object was detected. As seen in Fig. 11b, a pedestrian was successfully detected when our conversion model was used.

We next present a visual example for the PointRCNN model in Fig. 12. As seen in Fig. 12a and b, no object was detected even with and without our conversion model.

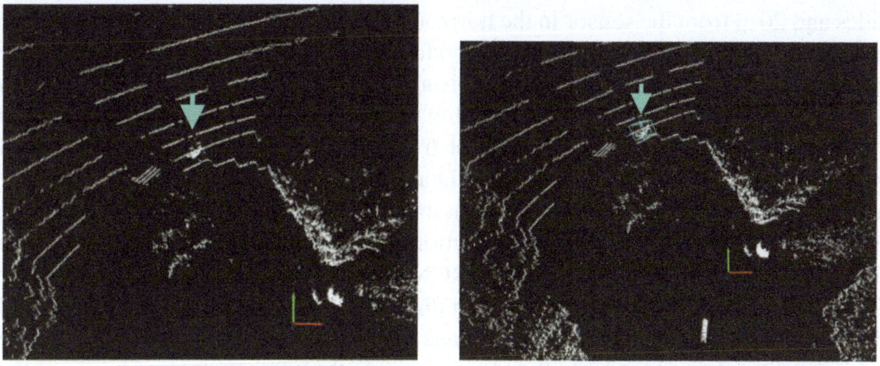

(a) Without our conversion model (b) With our conversion model

Fig. 11 Visual example of detection using Part-A2-free for evaluation data. The green bounding box indicates the detection result of a car, and the light blue bounding box indicates the detection result of a pedestrian. Distance of object was 6 m from sensor. Arrows indicate ground truth of pedestrian's position

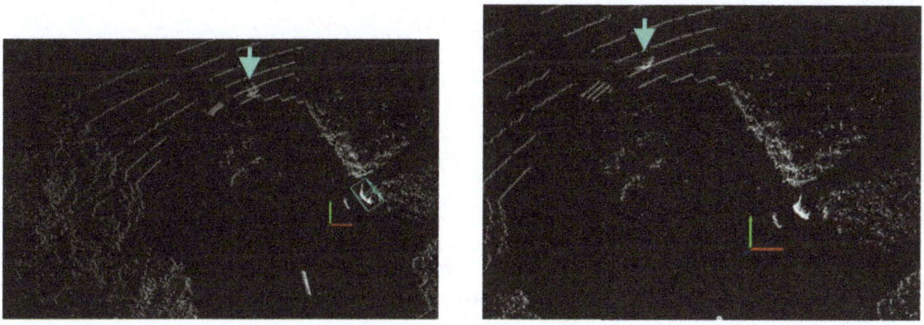

(a) Without our conversion model (b) With our conversion model

Fig. 12 Visual example of detection using PointRCNN for evaluation data. The green bounding box indicates the detection result of a car, and the light blue bounding box indicates the detection result of a pedestrian. Distance of object was 6 m from sensor. Arrows indicate ground truth of pedestrian's position

4.5 Detection Accuracy Versus Distance

We plot the detection accuracy of SECOND, Part-A2-Free, Part-A2-Anchor and PointRCNN with and without our conversion model in Figs. 13 and 14. The detection accuracy was plotted versus the distance from the sensor for angles of 0° and 15°. We used 100 frames of the evaluation data obtained from the experiment described in Sect. 4.1.4. The accuracy is defined as the ratio of the number of frames where detection is successful to the total number of frames (100). If a pedestrian is detected within 50 cm from the ground-truth position and only a pedestrian is detected at the position, the detection is considered successful. The range for detection is 30° in both sides and 20 m from the sensor in the horizontal plane. As seen in Fig. 13b, using our conversion model, the Part-A2-Anchor model improved detection accuracy when the distance was 6 m. However, when the distance was further than 6 m, the detection accuracy reduced drastically. As seen in Fig. 13c and d, PartA2-free and PointRCNN slightly improved detection accuracy for 6 m of distance. As seen in Fig. 14a and b, using our conversion model, SECOND and Part-A2-Anchor model successfully achieved higher detection accuracy when the distance was 6 m. However, when the distance was further than 6 m, the detection accuracy reduced drastically. As seen in Fig. 14c and d, PartA2-free and PointRCNN slightly improved detection accuracy for 6 m of distance. These results suggest that the conversion model partly improved detection accuracy. This was brought thanks to the training data produced with the conversion model, which considers loss of points through a window However, the data contained in the KITTI dataset were not acquired through a window. The existing models trained with such data do not fit object detection with loss of points.

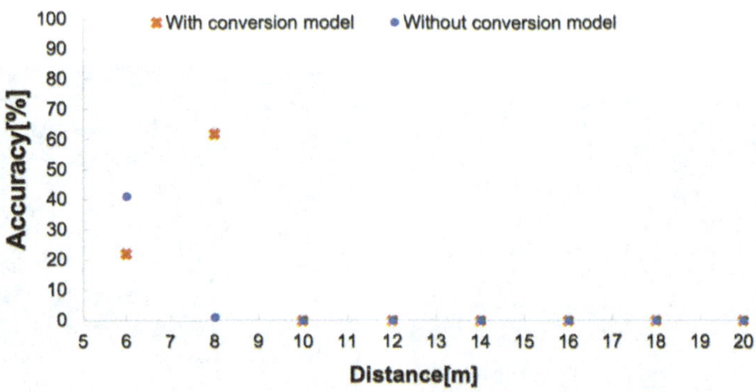

(a) SECOND

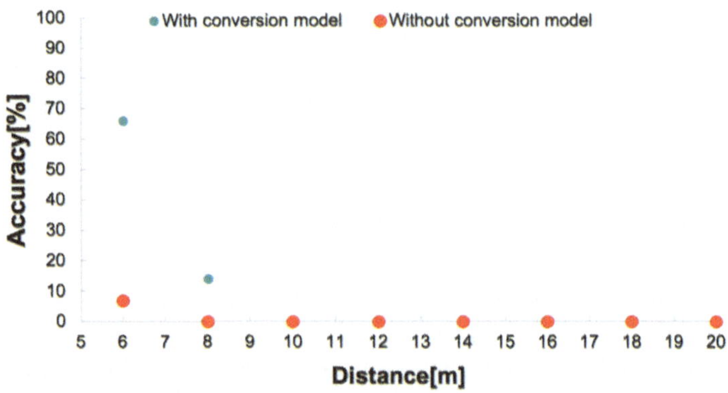

(b) Part-A2-Anchor

Fig. 13 Detection accuracy versus distance for angles of 0°

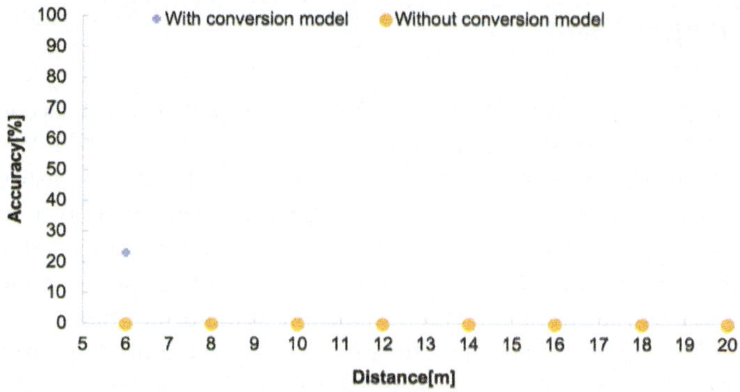

(c) Part-A2-free

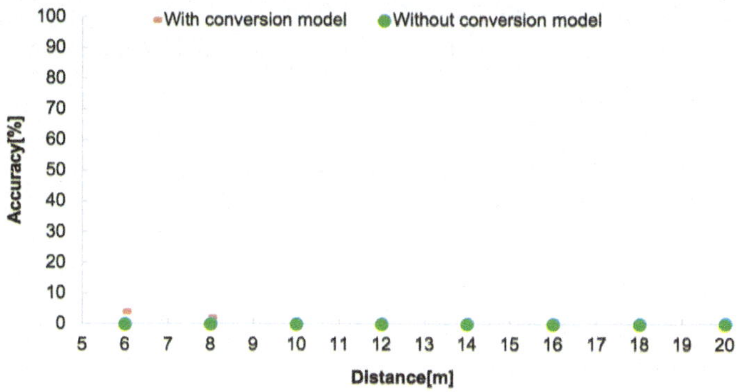

(d) Accuracy of pedestrian detection when using PointRCNN at 0 degree

Fig. 13 (continued)

5 Evaluation for Experiment at Real Intersection

This section describes the evaluation of the proposed system using data obtained at a real intersection.

5.1 Setup for Outdoor Experiments

We obtained data through a window using the same experiment system as the one described in Sect. 4.1.1. The experiment site was Gojyo-Shichihonmatsu intersection in Kyoto City, which is shown in Fig. 15. We applied MLMachine Learning models for detecting cars, pedestrians, and motorcycles to the data acquired at the experimental site. We used the same ML models as in Sect. 4.3: SECOND, PartA2-Anchor, PartA2-free, and PointRCNN.

5.2 Evaluation Result

We note that there is no grand-truth in these 3D data because the cameras could not be installed for privacy reasons. Figure 16 shows the result for SECOND. Without our conversion model, although multiple cars were detected, the directions of some of them were unnatural for the structure of the intersection. With our conversion model, the number of detected vehicles was smaller and the directions of the vehicles were more natural compared with the result obtained without our conversion model.

Figure 17 shows the result for PartA2-Anchor. Without our conversion model, although multiple cars were detected, the directions of some of them are unnatural for the structure of the intersection. With our conversion model, the directions of detected vehicles were more natural.

Figure 18 shows the result for PartA2-free. Without our conversion model, detection was not so accurate. Even with our conversion model, detection was not still accurate.

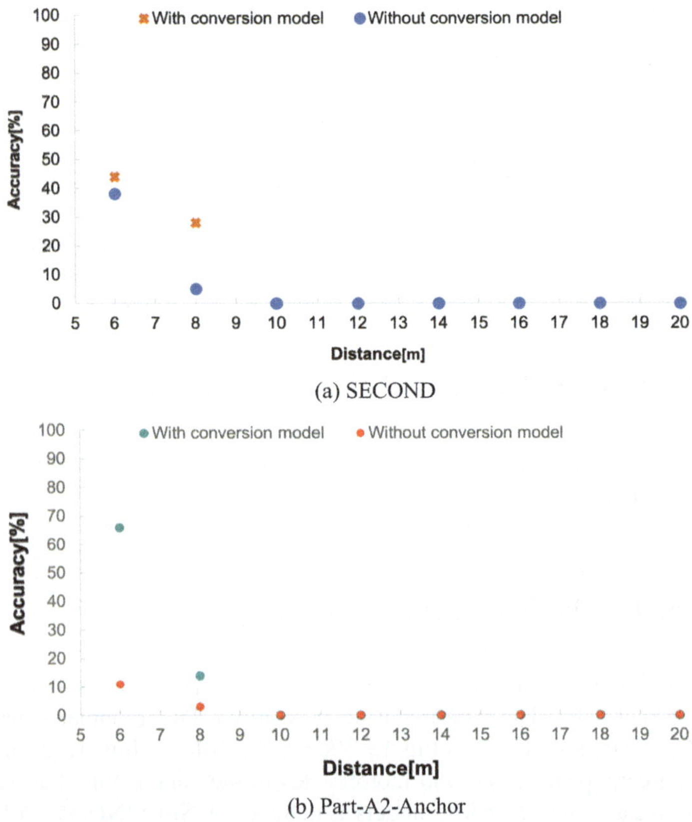

Fig. 14 Detection accuracy versus distance for angles of 15°

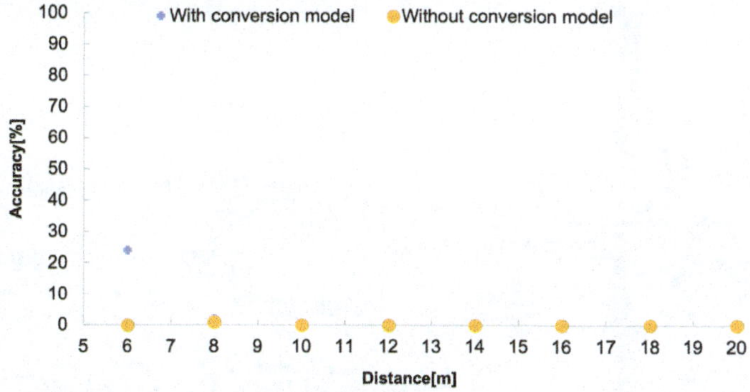

(c) Part-A2-free

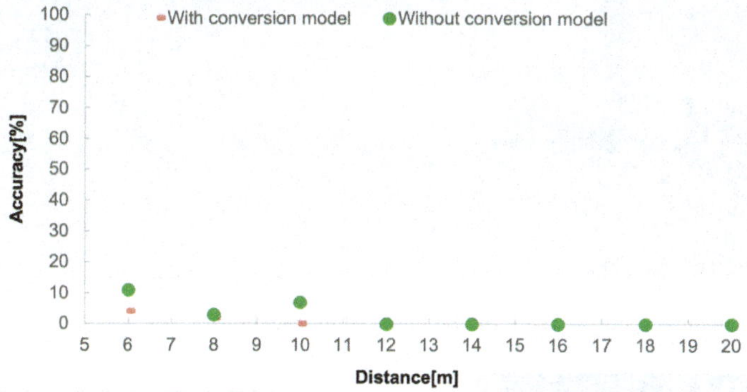

(d) Accuracy of pedestrian detection when using PointRCNN at 15 degrees

Fig. 14 (continued)

Figure 19 shows the result of detection for PointRCNN. No object was detected without using our conversion model. With our conversion model, a car was detected though its direction was not natural for the structure of the intersection.

6 Conclusion

To reduce the installation and maintenance costs of 3D sensor systems, this paper proposed a WFIWatch from inside system in which sensors deployed inside a building acquire 3D data through a window. The drawback of the system could be the deterioration in 3D data caused by the window. Therefore, we constructed a conversion model for data obtained without a window, so that the data can be used as training data for a ML model for object detection through a window. We constructed a conversion model for the proposed system through an experiment using a real LIDAR sensor device. Light detection and ranging From the result of the experiment, we verified

Fig. 15 Experimental scene at Gojo-Shichihonmatsu intersection

(a) Without our conversion model (b) With our conversion model

Fig. 16 Visualization of detection results using SECOND at intersection

that our conversion model performed improving detection accuracy. We also applied our conversion model to the data acquired at a real intersection. From the result, we confirmed that, although our conversion model is not always helpful for some of the MLMachine Learning models, it can determine the directions of detected vehicles more accurately than object detection without our conversion model. Future research includes construction of a conversion model with taking other parameters such as thickness of windows. The height dimension could be also considered to construct a 3D conversion model.

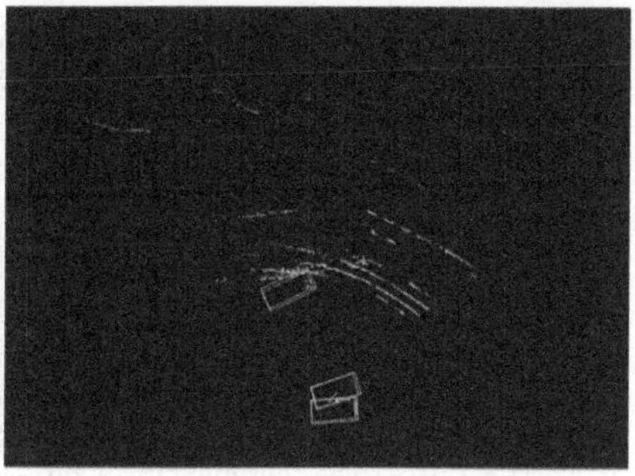

(a) Without our conversion model

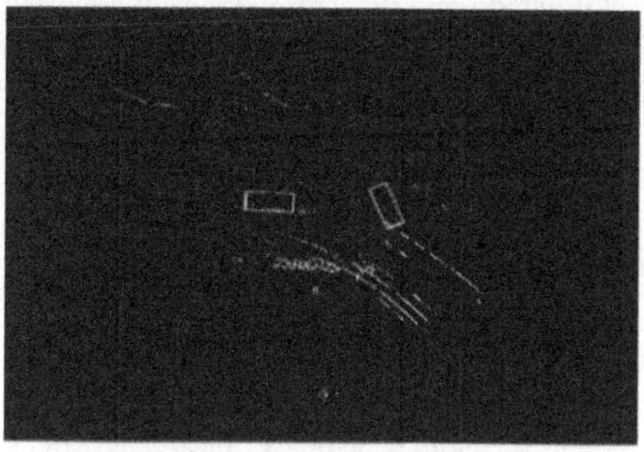

(b) With our conversion model

Fig. 17 Visualization of detection results using PartA2-Anchor at intersection

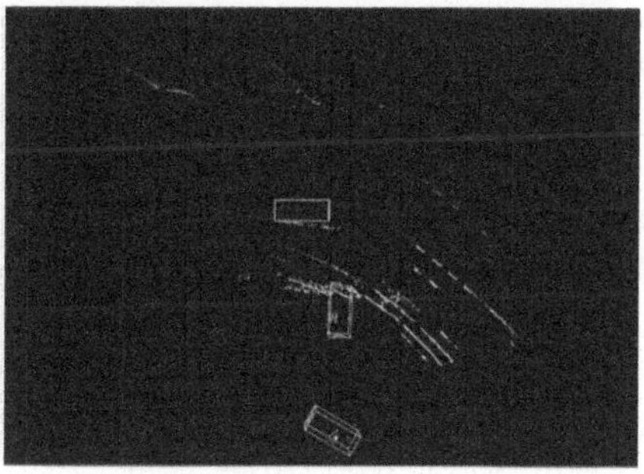

(a) Without our conversion model

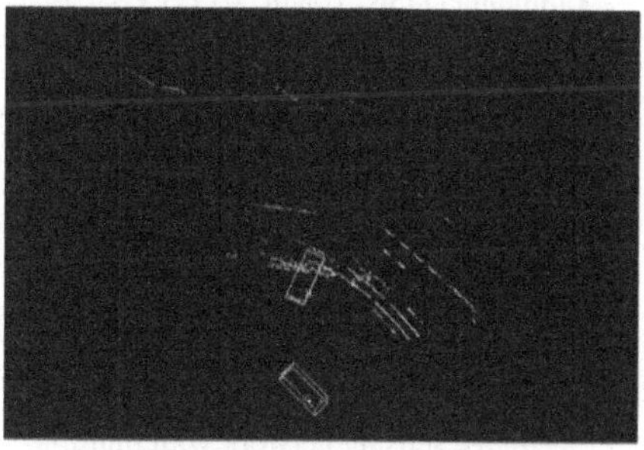

(b) With our conversion model

Fig. 18 Visualization of detection results using PartA2-free at intersection

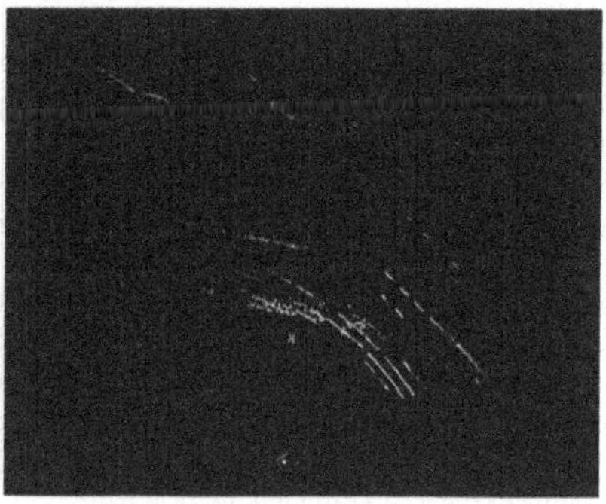

(a) Without our conversion model

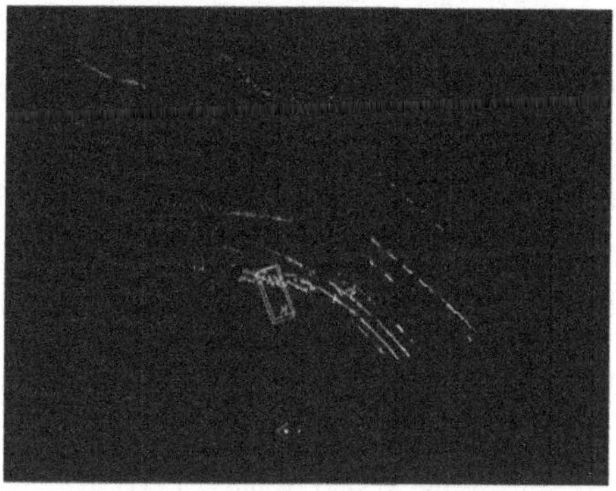

(b) With our conversion model

Fig. 19 Visualization of detection results using PointRCNN at intersection

Acknowledgements This work was supported in part by JSPS KAKENHI, Japan, under Grant no. 21H03427. These research results were partly obtained from the commissioned research (No.06401) by National Institute of Information and Communications Technology (NICT), Japan.

References

1. Jennifer B et al (2010) Getting clever about smart cities: new opportunities require new business models, vol 193. Cambridge, Massachusetts, USA, pp 244–277
2. Du R, Paolo S, Ming X, Vasilakos Athanasios V, Carlo F (2019) The sensable city: a survey on the deployment and management for smart city monitoring. IEEE Commun Surv Tutor 21(2):1533–1560
3. Hancke GP, Hancke GP Jr et al (2013) The role of advanced sensing in smart cities. Sensors 13(1):393–425
4. Emmanuel DSR, Yohan D, Peggy S, Pascal V (2016) A survey of vision-based traffic monitoring of road intersections. IEEE Trans Intell Trans Syst 17(10):2681–2698
5. Keiichiro S, Ryoichi S, Takehiro S, Eiji O, Takanori I, Dai K, Kozo S (2020) Prioritized transmission control of point cloud data obtained by lidar devices. IEEE Access 8:113779–113789
6. Akiyama K, Shinkuma R, Shiomi J (2022) Real-time adaptive data filtering with multiple sensors for indoor monitoring. In: NOMS 2022-2022 IEEE/IFIP network operations and management symposium. IEEE, pp 1–3
7. Li C, Shinkuma R, Sato T, Oki E (2021) Two-level processing scheme for 3d-image sensing network. In: 2021 IFIP networking conference (IFIP Networking). IEEE, pp 1–2
8. Chongyu L, Ryoichi S, Takehiro S, Eiji O (2021) Real-time data selection and merging for 3d-image sensing network with multiple sensors. IEEE Sens J 21(19):22058–22076
9. Strigel E, Meissner D, Seeliger F, Wilking B, Dietmayer K (2014) The ko-per intersection laser-scanner and video dataset. In: 17th International IEEE conference on intelligent transportation systems (ITSC), pp 1900–1901
10. Junxuan Z, Xu H, Hongchao L, Wu J, Yichen Z, Wu D (2019) Detection and tracking of pedestrians and vehicles using roadside lidar sensors. Trans Res Part C: Emerg Technol 100:68–87
11. Zhang ZY, Zheng JY, Wang X, Fan XL (2018) Background filtering and vehicle detection with roadside lidar based on point association. In: 2018 37th Chinese control conference (CCC), pp 7938–7943
12. Duarte F, António S, Rafael N, Cláudia S, Dibet G, Miguel G, Paulo N, João M, Pedro M-P (2021) Point-cloud based 3d object detection and classification methods for self-driving applications: a survey and taxonomy. Inf Fus 68:161–191
13. Yulan G, Hanyun W, Hu Q, Hao L, Li L, Mohammed B (2020) Deep learning for 3d point clouds: a survey. IEEE Trans Pattern Anal Mach Intell 43(12):4338–4364
14. Shi S, Wang X, Li H (2019) Pointrcnn: 3d object proposal generation and detection from point cloud. In: 2019 IEEE/CVF conference on computer vision and pattern recognition (CVPR), pp 770–779
15. Shi S, Wang Z, Shi J, Wang X, Li H (2020) From points to parts: 3d object detection from point cloud with part-aware and part-aggregation network. IEEE Trans Pattern Anal Mach Intell
16. Shi S, Guo C, Jiang L, Wang Z, Shi J, Wang X, Li H (2020) Pv-rcnn: point-voxel feature set abstraction for 3d object detection. In: Proceedings of the IEEE/CVF conference on computer vision and pattern recognition, pp 10529–10538
17. Zhou Y, Tuzel O (2018) Voxelnet: End-to-end learning for point cloud based 3d object detection. In: Proceedings of the IEEE conference on computer vision and pattern recognition, pp 4490–4499

18. Lang AH, Vora S, Caesar H, Zhou L, Yang J, Beijbom O (2019) Pointpillars: fast encoders for object detection from point clouds. In: Proceedings of the IEEE/CVF conference on computer vision and pattern recognition, pp 12697–12705
19. Yan Y, Yuxing M, Bo L (2018) Second: sparsely embedded convolutional detection. Sensors 18(10):3337
20. Velodyne. Puck lidar sensor, high-value surround lidar|velodyne lidar. https://velodynelidar.com/products/puck/. Accessed 27 Sept 2021
21. NVIDIA. Getting started with jetson nano developer kit. https://developer.nvidia.com/embedded/learn/get-started-jetson-nano-devkit. Accessed 27 Sept 2021
22. Hall CA, Weston Meyer W (1976) Optimal error bounds for cubic spline interpolation. J Approx Theory 16(2):105–122
23. SciPy documentation. https://docs.scipy.org/doc/scipy/index.html. Accessed 15 July 2022
24. Geiger A, Lenz P, Stiller C, Urtasun R (2013) Vision meets robotics: the kitti dataset. Int J Robot Res (IJRR)
25. OpenPCDet Development Team (2020) Openpcdet: an open-source toolbox for 3d object detection from point clouds. https://github.com/open-mmlab/OpenPCDet

Federated Learning with Client Selection in Resource-Uncertain Wireless Networks: Simulation and Proof of Concept Experiments

Naoya Yoshida, Takayuki Nishio, Koji Yamamoto, and Masahiro Morikura

Abstract Federated learning (FL) is an emerging machine learning (ML) framework that enables the use of often privacy-sensitive smart city data (e.g., pictures from the road surveillance cameras and connected vehicles) for machine learning. However, in the smart city scenario, data-holder devices called clients are heterogeneous, and their computation and communication resources available for FL are uncertain and fluctuating. This paper presents a multi-armed bandit (MAB)-based client selection method for FL which does not require prior knowledge of the computation and communication resources of clients. The MAB-based method achieves a better tradeoff between exploration and exploitation and reduces the time consumption for model updating. Simulation results demonstrated that the proposed scheme requires less learning time than that required by the conventional method in a resource-fluctuating scenario. Furthermore, experimental results using the off-the-shelf devices demonstrated the feasibility of FL and MAB-based client selection in a real wireless network environment.

Acronyms

FL	Federated Learning
ML	Machine Learning
MAB	Multi-armed Bandit
IoT	Internet of Things
IID	Independent Identical Distribution
LAN	Local Area Network
FedCS	FL with Client Selection

N. Yoshida · K. Yamamoto · M. Morikura
Graduate School of Informatics, Kyoto University, Kyoto, Japan

T. Nishio (✉)
School of Engineering, Tokyo Institute of Technology, Tokyo, Japan
e-mail: nishio@ict.e.titech.ac.jp

R. Shinkuma et al. (eds.), *Advances in Engineering and Information Science Toward Smart City and Beyond*, Engineering Cyber-Physical Systems and Critical Infrastructures 5, https://doi.org/10.1007/978-3-031-29301-6_11

BS	Base Station
RB	Resource Blocks
UCB	Upper Confidence Bound
CNN	Convolutional Neural Network
LTE	Long-term Evolution
AP	Access Point

Glossary

Access point	is a device that acts as a central hub for wireless communication, allowing wireless devices such as laptops, smartphones, and tablets to connect to the network.
Arm	refers to the different options or choices available to the decision-maker.
Client	is a device or local server that participates in the FL training process.
Client selection	is one of the procedures in FL to select a client to participate in that round among many clients.
Federated learning	is a machine learning framework where a model is trained on decentralized data distributed across multiple devices, such as on users' phones or hospitals' data servers. It allows for training machine learning models without the need to transfer the data to a centralized server.
Global model	is a machine learning model that is trained by aggregating the model updates from multiple clients.
Local model	is a machine learning model that is trained by clients using their own data.
Multi-armed bandit	is a machine learning problem in which an agent must choose between several options called "arms" to maximize its reward. Each armis associated with a certain reward probability, and the agent must balance the exploration of different arms in order to learn about their reward probabilities and the exploitation of the most rewarding arms in order to maximize its overall reward.
Reward	refers to the benefit or value obtained by choosing a particular arm.
Round	is one iteration of the FL training process, where the clients participate by locally training a model using their own data, and then sending the updated model parameters to a central server.

1 Introduction

Internet of things (IoT) combined with machine learning (ML) has the great potential to create smarter and novel smart city applications for making our daily life more comfortable. However, the challenge is how to utilize IoT data which often requires privacy considerations. Emerging ML technology, federated learning (FL), is gaining a lot of attention; which is a collaborative ML framework that enables model training in a privacy-preserving manner using the rich data and computational resources of the massive IoT devices [1, 2]. FL iteratively asks data-holder devices (e.g., gateways, edge servers, smart mobility, and smart devices) to (1) obtain a trainable model from a certain server, (2) update the model with their data, and (3) upload the updated model to the server. Then, the server (4) aggregates the models to generate a further improved model. This method is unlike the conventional ML methods that acquire and store raw data in central systems, which have the risk of exposing the private and/or sensitive data of mobile users. FL can be applied to model training tasks that require data may containing privacy-sensitive information owing to the privacy advantage. Its typical applications, such as word and emoji prediction on a keyboard [3] and environment monitoring [4], have been summarized in [5].

Latency caused by wireless transmission and local computation on client devices is a significant problem in FL, particularly when the number of participating clients in the FL process is significant [6, 7]. Therefore, latency in FL must be reduced to realize its practical applications. However, there is a time constraint on FL with smartphones and other mobile devices. This constraint necessitates that training processes be performed at nonpeak hours, such as at midnight or early morning when the network is not congested because the ML models to be trained and communicated with are typically large. Furthermore, FL must operate with low power consumption. The longer the FL process, the more power is consumed; therefore, we must reduce the learning time of the FL. The FL latency increases because mobile clients with different wireless-link qualities and computational capabilities must update and transmit a several-gigabyte ML model. A direct application of the FL algorithm without any consideration of such heterogeneous client properties will make the overall training process inefficient. For instance, when some clients have limited computational resources, they will require a longer time to update the models. Moreover, when the clients are operating under poor wireless channel conditions, the update time is even longer.

1.1 Related Works and Problem Statement

Recently, some studies have investigated the major problems related to the implementation of FL in wireless networks. The works in [7–10] provided a comprehensive review of the designs of FL algorithms and focused on various challenges, problems, and solutions to enhance the effectiveness of FL.

One approach to improve the FL performance is algorithm optimization, which exchanges the model parameters less frequently by performing more computations on the clients to reduce the communication cost [1]. This significantly reduces the network traffic when compared to an algorithm iterating local model update and model parameter exchange alternately [11]. In [12], the authors proposed communication-efficient FedAvg (CE-FedAvg), which reduced the number of communication rounds by optimizing ML models using an Adam optimizer and model compression. Another study, [13], minimized the communication cost by evolving the structure of neural networks using a modified sparse evolutionary training method.

Model compression is another widely studied approach to reduce the amount of data shared between the server and clients. A study, [14], proposed a new compression framework that compressed both the upstream and downstream communications and helped achieve a high-accuracy model in situations where the client data follow non-independent identical distributions (non-IID) or only a small fraction of clients participate in every communication round.

Another important approach to reduce the training latency is "client selection", which limits the number of participating mobile clients by selecting specific users [15, 16]. A challenge of client selection is how to efficiently obtain a set of clients that can improve the accuracy with low latency whereas the accuracy improvement is difficult to estimate until the training is conducted. Nishio and Yonetani have proposed a simple client selection algorithm that increases the number of participating devices and improves accuracy by selecting devices with the least increase in computation and communication latency [15]. In [17], a new learning mechanism extending FL and a client-selection algorithm considering data non-IIDness were proposed. A client-scheduling algorithm that jointly considers the learning aspect (i.e., model improvement) and communication aspect (i.e., instantaneous wireless channel quality) was studied [18], and a mechanism to jointly control wireless resources and client scheduling has been proposed [19].

The aforementioned client-selection algorithms assume that accurate resource information (e.g., throughput, channel state, and available computation power of clients) is available, and the amount of resources in a round is stable. However, actual computation and communication resources are uncertain and fluctuate owing to unstable wireless links, network congestion, and background computation tasks. Therefore, it is difficult to obtain accurate resource information a priori. Consequently, the existing selection strategy based on the ideal assumptions of resource information can suffer from significant performance degradation induced by the gap between assumption and practice.

1.2 Our Work

To address this problem, we propose a reinforcement-learning-based client-selection algorithm to improve client selection in an environment, wherein the amount of client resources cannot be known a priori and changes in every round. The proposed algorithm uses a multi-armed bandit (MAB) algorithm [20] to achieve an improved tradeoff between exploration (which aims to estimate the expected amount of the client resource) and exploitation (which aims to leverage the resource-rich clients) to minimize the latency. The MAB problem is a sequential decision problem, which is typified by a slot-machine game, wherein the player selects an arm of a slot machine in each time slot and observes the reward determined probabilistically for the selected arm. The objective of the MAB algorithm is to determine the arm to maximize the cumulative reward obtained in sequential selections. In the client-selection problem, the arms correspond to the set of clients, and the reward is a value that can evaluate the efficiency of FL, such as the learning time.

The application of MAB to resource scheduling in wireless networks has been widely studied [21, 22]. One advantage of MAB algorithms is their computation efficiency; therefore, the resource operator can determine scheduling in a short duration. This fits the low-latency requirement of the client-selection algorithm. However, how to apply the MAB problem to the FL client selection for heterogeneous clients in mobile networks or whether the MAB-based client selection works effectively for the tasks is not clear.

Therefore, we focus on applying MAB to the client-selection problem and demonstrate its effectiveness through computer simulations and experiments using off-the-shelf devices. The contributions of this study are summarized as follows:

- We propose a client selection method that incorporates MAB for FL in smart city scenarios, wherein client resources are heterogeneous, uncertain, and fluctuating. We designed two MAB-based client-selection methods: one is the straightforward application of MAB to client selection, and the other is the proposed MAB-based client-selection method that enables a significant reduction in the number of arms and improve the convergence of the client-selection algorithm. The simulation results show that the proposed method can achieve shorter learning latency than that of the existing methods when the client resources highly fluctuate.
- We experimentally demonstrate the feasibility of FL with the proposed client selection. We implement the FL application on several single-board computers with GPU (used as FL clients) and MacBooks (used as FL clients and an FL server) and connect them over an IEEE 802.11ac wireless local area network (LAN). The evaluation of the actual devices enabled us to investigate the effect of resource fluctuations as distinct from the artificially generated resource fluctuation in simulations.

This is an extended paper of our previous paper [23] by including more detailed explanations and discussion, and an experimental feasibility study. In parallel with and independent of this study, a similar concept of adapting the MAB algorithm to the client selection in FL has been studied in [24, 25]. Cho et al. [24] focused on achieving faster convergence in training by the biased selection based on local loss and did not consider latency increase caused by the resource heterogeneity, unlike our work. On the other hand, [25] considers the bottleneck caused by the computation and communication resource heterogeneity and minimizes the training latency, including computation and communication latency, as in our work. In [25], it is assumed that the number of selected clients is less than or equal to that of available wireless channels to ignore the inter-client interference; however, the algorithm proposed by us has no limit on the number of clients owing to assumptions on the wireless environment. Additionally, we performed a more large-scale evaluation (with more clients and more difficult ML tasks), including the experimental evaluations.

2 Prior Work: Federated Learning with Client Selection

In this section, we introduce the FL with client selection (FedCS), as presented in [15]. Algorithm 1 describes the FedCS. The objective of FedCS is to work with heterogeneous clients in actual mobile networks without causing large training latency induced by some bottleneck devices with limited computation power or poor wireless channel. In FedCS, (1) the server first initializes a global model with randomized weight. Subsequently, the following steps are executed iteratively. (2) The server sends a request to $\lceil K \times C \rceil$ clients, which are randomly chosen, to send resource information (i.e., current wireless-channel states and computational capacities) to the FL server, where K denotes the total number of clients, $C \in (0, 1]$ a hyperparameter representing the proportion of clients participating in each round to the total number of clients, and $\lceil \cdot \rceil$ the ceiling function. (3) The server selects clients that proceed to the subsequent steps based on the client selection algorithm shown in Algorithm 2 explained in Sect. 3.2.1. (4) The server then distributes the global model to the selected clients. (5) The clients which received the global model update the models with their own data. (6) Then, the updated parameters are uploaded to the server. (7) Finally, the server aggregates the obtained models updated by the selected clients to generate an improved global model.

Algorithm 1 Federated Learning with Client Selection. Parameter K denotes the number of clients, and $C \in (0, 1]$ denotes the fraction of random clients that receive a resource request in each round.

1: `Initialization:` The server first initializes a global model either randomly or by pretraining using public data.
2: `Resource Request:` The FL server then asks $\lceil K \times C \rceil$ random clients to participate in the current training task. The clients who receive the request notify the operator about their data and resource information.
3: `Client Selection:` Using this information, the FL server determines which of the clients will proceed to the subsequent steps to complete the steps within a certain deadline.
4: `Distribution:` The server distributes the parameters of the global model to the clients selected to update the models locally.
5: `Model Update:` The selected clients update the models using their data.
6: `Scheduled Upload:` The selected clients upload the new parameters using the resource blocks (RBs) allocated by the FL server.
7: `Aggregation:` The server then averages the updated parameters, after which it replaces the global model with the averaged model.
8: All steps except `Initialization` are iterated multiple times until the global model achieves the desired performance, or the final deadline arrives.

3 Federated Learning with Multi-armed Bandit-Based Client Selection

3.1 System Model

The system model is based on a previous work [15]. We consider a mobile network composed of a base station (BS) and mobile clients, and an FL server is deployed at the network edge. The FL server determines which clients participate in the FL task.

The mean of path loss between the BS and each client is stable because the mobile clients are assumed not to move. We assume that a fixed number of RBs [26] are available at the BS. The RBs are shared among the clients; therefore, when multiple clients transmit their local models simultaneously, the RBs are equally allocated to the clients, and the throughput for each client reduces compared to a case where one client transmits its model solely. We also assume that the modulation and coding schemes for radio communications are determined suitably for each client so that the throughput of the link is maximized. These lead to different throughputs for each client in each round, while the mean throughput for a client is stable during the FL task. Each client throughput is also affected by the background traffic on the client. Although the throughput is determined by a combination of these factors, for simplicity, we assume that the throughput and resulting upload time of the local model follow unknown distributions that do not change during the FL task.

In addition, we assume that the available computation power of each client in each round for the FL task also fluctuates due to the background computation load on the client. In FL, the participating clients are user equipment such as smartphones and tablets; they have computation tasks not related to FL. Therefore, the available computation power for FL can fluctuate. We assume that the mean computation power in each round and the resulting model updating time follow unknown distributions as that of communication resources.

In this paper, the realizations of throughput $\theta_{k,r}$ and computation power $\delta_{k,r}$ measured at client k are called resource information in the rth round. Model upload time $t_{k,r}^{\mathrm{UL}}$ and model update time $t_{k,r}^{\mathrm{UD}}$ are determined by $\theta_{k,r}$ and $\delta_{k,r}$, respectively. The FL server can obtain the resource information of a client only when the client is selected and performs the model update and upload.

3.2 Multi-armed Bandit-Based Client Selection

3.2.1 Existing Client-Selection Algorithm and Its Extension

This subsection firstly describes the original FedCS algorithm [15] and then its extension.

At the `Client Selection` step in FedCS, the FL operator selects the maximum possible number of clients that have sufficient computation power and communication bandwidth to complete the `Distribution`, `Model Update`, and

`Scheduled Upload` steps within a certain deadline. The client selection strategy is formulated by the following maximization problem [15]

$$\begin{aligned} &\underset{S \subset K}{\text{maximize}} \quad |S| \\ &\text{subject to} \quad T_{\text{round}} \geq T_{\text{CS}} + T_S^{\text{d}} + T_S^{\text{c}} + T_{\text{agg}}, \end{aligned} \tag{1}$$

where $\boldsymbol{K} = \{1, \ldots, K\}$, $\boldsymbol{K}' \subset \boldsymbol{K}$, and $\boldsymbol{S} \subseteq \boldsymbol{K}'$ denote index sets of all clients, randomly chosen clients in the `Resource Request` step, and selected clients, respectively. $|\boldsymbol{K}'| = \lceil K \times C \rceil$. T_{round} denotes a latency threshold for a round, and T_{CS}, T_S^{d}, T_S^{c}, and T_{agg} denote latency for selecting clients, distributing models to clients, total latency for clients' model updates and uploads, and aggregating uploaded models, respectively. This objective was inspired by the result from [1] that a larger number of clients participated in each round reduces the training latency for global models to achieve certain performance thresholds.

To solve this optimization problem, FedCS uses a greedy algorithm that adds a client k with the least increase in latency $T_{\text{inc}}(\boldsymbol{S}, k)$ to $\boldsymbol{S}$ one by one. $T_{\text{inc}}(\boldsymbol{S}, k)$ is defined as the estimated latency increase, which indicates how much the round time will increase when adding client k to $\boldsymbol{S}$. Since the computation power of the server is generally much larger than that of clients, T_{CS} and T_{agg} can be negligible. Thus, we should estimate the latency for distributing, updating, and uploading models to estimate $T_{\text{inc}}(\boldsymbol{S}, k)$. Let $T_S^{\text{d}} = \max_{k_i \in \boldsymbol{S}}\{t_{k_i,r}^{\text{UL}}\}$ be the estimated time required for the `Distribution` step in the rth round. $T_{\text{inc}}(\boldsymbol{S}, k)$ and T_S^{d} are determined using $t_{k,r}^{\text{UL}}$ and $t_{k,r}^{\text{UD}}$, which are latency for updating and uploading a model by client k for each round, respectively. For simplicity, r is omitted in the following. $T_{\text{inc}}(\boldsymbol{S}, k)$ is given by

$$T_{\text{inc}}(\boldsymbol{S}, k) = \left(T_{\boldsymbol{S}\cup k}^{\text{d}} - T_S^{\text{d}}\right) + \max\{t_k^{\text{UD}} - (t - T_S^{\text{d}}), 0\} + t_k^{\text{UL}}. \tag{2}$$

$$T_S^{\text{d}} = \max_{k_i \in \boldsymbol{S}} t_{k_i}^{\text{UL}} \tag{3}$$

Algorithm 2 Client selection.

Require: Index set of randomly selected clients $\boldsymbol{K}'$
1: **Initialization** $\boldsymbol{S} \leftarrow \{\}, t \leftarrow 0$
2: **while** $|\boldsymbol{K}'| > 0$ **do**
3: $\quad x \leftarrow \arg\max_{k \in \boldsymbol{K}'} f(\boldsymbol{S}, k)$
4: $\quad$ remove x from $\boldsymbol{K}'$
5: $\quad t' \leftarrow t + T_{\text{inc}}(\boldsymbol{S}, x)$
6: $\quad$ **if** $|\boldsymbol{S}| < S_{\text{round}}$ **then**
7: $\qquad t \leftarrow t'$
8: $\qquad$ add x to $\boldsymbol{S}$
9: $\quad$ **end if**
10: **end while**
11: **return** $\boldsymbol{S}$

As mentioned above, FedCS [15] maximizes the number of selected clients while satisfying a certain deadline. However, the algorithm can be easily extended to one that minimizes the learning time while satisfying a certain number of selected clients. The extended algorithm is given in Algorithm 2. Here, $f(\boldsymbol{S}, k)$ denotes the importance of the client (i.e., how much priority should be given to selecting the client.) In the extended FedCS algorithm, we use $-T_{\text{inc}}(\boldsymbol{S}, k)$ for $f(\boldsymbol{S}, k)$. Thus, the extended FedCS iteratively selects a client with the least latency, and, subsequently, updates $\boldsymbol{S}$ until the number of selected clients $|\boldsymbol{S}|$ reaches the required number S_{round}.

As mentioned in Sect. 1, the performance of FedCS or the extended FedCS degrades when client resources fluctuate because the algorithm needs prior information about the client resources. The actual latency for updating and uploading models can be different from $T_{\text{inc}}(\boldsymbol{S}, k)$ estimated using resource information obtained in `Resource Request` because of the resource fluctuation. Therefore, an FL round frequently consumes more time than $T_{\text{inc}}(\boldsymbol{S}, k)$ because of $\max\{\cdot\}$ in (2).

3.2.2 Naive Application of MAB to Client Selection

This paper presents a client selection algorithm that works especially well when client resources fluctuate and cannot be known until the clients are selected. The MAB algorithm typified by the upper confidence bound (UCB) policy [20] is adopted for balancing exploration (which selects clients that are selected less often) and exploitation (which selects clients with rich resources) and achieving higher accuracy with low training latency.

We first consider a naive application of MAB. The application scenario of MAB is typified by a slot machine selection, in which a player selects an arm of the slot machine from multiple slots. The reward is obtained by selecting an arm, and the player updates the arm selection strategy based on the reward to maximize the cumulative reward. The key challenge of MAB is to balance the tradeoff between exploration and exploitation. It is necessary to select each arm a sufficient number of times to accurately estimate the expected reward for each arm. However, selecting an arm with a low expected reward during the exploration will reduce the cumulative reward. On the other hand, exploitation, i.e., selecting the arm with the largest estimated expected reward at a certain time, could miss the better arm and not obtain the largest cumulative reward. To achieve a better tradeoff between exploration and exploitation, the UCB algorithm calculates the following UCB score and selects an arm with the maximal UCB score;

$$\overline{\mu_i} + \sqrt{\frac{\log \sum_i N_i}{2N_i}}, \tag{4}$$

where i denotes the index of an arm, $\overline{\mu_i}$ denotes the sample mean of the ith arm reward $\mu_i \in [0, 1]$, and N_i denotes the number of times the ith arm has been selected. The UCB algorithm guarantees the upper bound of regret, which is the cumulative difference between the reward obtained by this algorithm and the expected reward obtained by the optimal policy.

We apply the UCB algorithm to the client selection naively. In the client selection, an arm and reward in MAB can be defined as a set of clients participating and the inverse of learning latency in an FL round, respectively. However, this naive application of the UCB algorithm will struggle to converge because the number of arms will be significantly large. When we select $S_{\mathrm{round}} = 10$ clients from $\lceil K \times C \rceil = 30$ clients, the number of arms is $\binom{\lceil K \times C \rceil}{S_{\mathrm{round}}} > 30$ million.

3.2.3 Proposed MAB-Based Client Selection

To avoid such a combinatorial explosion, we propose a novel MAB-based client selection algorithm, which regards a client as an arm and creates a set of clients by adding each client sequentially. This selection method allows us to reduce the number of arms to be the same as the number of clients (i.e., K), and in each round, the arms are filtered to be the same number of randomly chosen clients (i.e., $\lceil K \times C \rceil$).

The proposed MAB-based client selection is based on the extended FedCS detailed in Algorithm 2. The selection of one client corresponds to lines 3–9 in Algorithm 2 using UCB score for $f(\boldsymbol{S}, k)$. The FL server calculates the UCB score we newly define for each $\lceil K \times C \rceil$ client, and S_{round} clients are selected. We define two UCB scores for $f(\boldsymbol{S}, k)$: naive UCB score and element-wise UCB score.

The naive UCB score is defined as follows. We use measured latency increase by adding client k to $\boldsymbol{S}$, $T'_{\mathrm{inc}}(\boldsymbol{S}, x)$ as reward. $T'_{\mathrm{inc}}(\boldsymbol{S}, x)$ is estimated from the model update and upload times measured in the `Scheduled Upload` step. Let $\overline{T_{\mathrm{inc}}}(\boldsymbol{S}, k)$ be the sample mean of $T'_{\mathrm{inc}}(\boldsymbol{S}, x)$. The naive UCB score used to $f(\boldsymbol{S}, k)$ is given by

$$f(\boldsymbol{S}, k) = \frac{-\overline{T_{\mathrm{inc}}}(\boldsymbol{S}, k)}{\alpha} + \sqrt{\frac{\log \sum_{i=1}^{|\boldsymbol{S}|} N_k}{2N_k}}, \tag{5}$$

where N_k denotes the number of times that client k has been selected. The value of constant weight α is set to normalize the sample mean of latency increase for balancing the first and second terms. In the naive UCB score, the less frequently a client is selected, the larger the second term for the client is. Therefore, clients that are less frequently selected are more likely to be selected, even if the first term for the clients is large (i.e., the expected latency increase is large at the time.) The aim of this naive UCB score is to directly reduce the round time by utilizing the UCB score setting. However, convergence with this UCB score is not guaranteed because the UCB score depends on the UCB scores of other clients.

The element-wise UCB score is defined as

$$f(\boldsymbol{S}, k) = -T'_{\text{inc}}(\boldsymbol{S}, k), \tag{6}$$

$$= -\left(T^{\text{d}}_{\boldsymbol{S}\cup k} - T^{\text{d}}_{\boldsymbol{S}}\right) - \max\{\tau_k^{\text{UD}} - (t - T^{\text{d}}_{\boldsymbol{S}}), 0\} - \tau_k^{\text{UL}}, \tag{7}$$

$$T^{\text{d}}_{\boldsymbol{S}} = \max_{k_i \in \boldsymbol{S}} \tau_{k_i}^{\text{UL}}, \tag{8}$$

$$\tau_k^{\text{UD}} = \frac{\overline{t_k^{\text{UD}}}}{\beta} - \sqrt{\frac{\log \sum_{i=1}^{|\boldsymbol{S}|} N_k}{2N_k}}, \tag{9}$$

$$\tau_k^{\text{UL}} = \frac{\overline{t_k^{\text{UL}}}}{\gamma} - \sqrt{\frac{\log \sum_{i=1}^{|\boldsymbol{S}|} N_k}{2N_k}}, \tag{10}$$

where $\overline{t_k^{\text{UL}}}$ and $\overline{t_k^{\text{UD}}}$ denote the sample means of the model upload and update times of client k, respectively. Constant weights β and γ are set to normalize the model upload and update time. τ_k^{UD} and τ_k^{UL}, respectively, denote the model update and upload times with negative UCB amendment terms. In contrast to the naive UCB score, the second terms in (9) and (10) are negative. Thus, the fewer times a client k is selected, the smaller τ_k^{UD} and τ_k^{UL} will be. Consequently, clients with a small number of selections are more likely to be selected, even if the mean values of the model upload and update times of the client are relatively long, for exploitation. Since τ_k^{UD} and τ_k^{UL} are determined independently for each client, the UCB scores are not affected by other arms, as assumed in the UCB algorithm.

4 Simulation and Experimental Evaluation

We performed a simulation evaluation using large-scale datasets and a proof of concept experiment using off-the-shelf Wi-Fi devices. The simulation evaluation indicated that the proposed algorithm reduced the total time for the FL task. The experimental evaluation demonstrated the feasibility of FL with the proposed client-selection algorithm.

4.1 ML Setups

We adopted an image classification task using publicly available large-scale image datasets in the simulated communication networks. We used the CIFAR-10 dataset [27], which is a 10-class image classification dataset consisting of 50,000 training samples and 10,000 testing samples. The CIFAR-10 dataset has often been used as a benchmark task in FL studies [1, 28]. The training samples were separated and distributed to $K = 100$ clients as follows. The number of images owned by each client was randomly determined in the range of 100–1000. Subsequently, we split the training dataset based on the IID settings, in which each client randomly sampled a specified number of images from the entire training dataset.

A convolutional neural network (CNN), which is often used for image classification tasks, was used. The architecture of the CNN was the same as that in the previous work [15], which consisted of six convolutional layers and three fully connected layers. The convolution layers had 32, 32, 64, 64, 128, and 128 channels, each of which was activated using ReLU and then batch normalized, and every two of which were followed by 2×2 max pooling. The two fully connected layers had 512 and 192 units with ReLU activation, respectively, and another had 10 units activated using softmax. The CNN had approximately 4.6 million parameters; thus $M = 18.3$ MB in 32-bit float. As in [15], we used SGD optimizer with 0.25 for the initial learning rate and 0.99 for learning-rate decay, and mini-batch size and the number of epochs in each round were 50 and 5, respectively.

4.2 Simulation Settings

The simulation setting was based on [15]. An edge FL server and a BS were located at the center of a 2-km radius cell of long-term evolution (LTE) network, and $K = 100$ clients were uniformly distributed in the cell. The frequency and channel model for the LTE network were 2.5 GHz and an urban channel model defined in the ITU-R M.2135-1 Micro NLOS model with a hexagonal cell layout [29], respectively. The antenna heights of the BS and clients were 11 m and 1 m, respectively, and the transmission power and antenna gain of the BS and clients were 20 dBi and 0 dBi. We assumed that 10 RBs, corresponding to a bandwidth of 1.8 MHz, were available for the FL in each time slot of 0.5 ms. To calculate the throughput, we used the Shannon capacity with a certain loss with $\Delta = 1.6$ and $\rho_{\max} = 4.8$, as used in [30]. In this setting, the maximum and mean throughputs of the clients were 8.6 and 1.4 Mbit/s, respectively. These values are realistic for an LTE network.

For simulating throughput fluctuation, we utilized the aforementioned throughput model to obtain the average throughput of each client θ_k, and then the throughput in each round was sampled from a truncated normal distribution with a mean of θ_k. Let $\mathbb{R}$ be a set of real numbers ($\mu, \sigma, a, b \in \mathbb{R}$), and $a \leq \mu \leq b$. A cumulative distribution function of the truncated normal distribution for $a \leq x \leq b$ is given by

$$F(x; \mu, \sigma, a, b) = \frac{\Phi\left(\frac{x-\mu}{\sigma}\right) - \Phi\left(\frac{a-\mu}{\sigma}\right)}{\Phi\left(\frac{b-\mu}{\sigma}\right) - \Phi\left(\frac{a-\mu}{\sigma}\right)}, \tag{11}$$

where $\Phi(\cdot)$ denotes the cumulative distribution function of the standard normal distribution. $\Phi(\cdot)$ is given by

$$\Phi(x) = \frac{1}{2}\left(1 + \mathrm{erf}\left(x/\sqrt{2}\right)\right). \tag{12}$$

Let $\eta \in \mathbb{R}, \eta < 2$ be a parameter that denotes the intensity of resource fluctuation. In the simulation, we sampled the throughput from the truncated normal distribution in (11), where $\mu = \theta_k,\ \sigma^2 = \theta_k^\eta,\ a = \theta_k - \sigma,\ b = \theta_k + \sigma$. Therefore, as increasing η, the throughput fluctuated more largely.

For the computational capability, we assumed that the server has sufficiently high computation power to ignore the latency to conduct client selection and model aggregation (i.e., $T_{\mathrm{CS}} = T_{\mathrm{agg}} = 0$), whereas that of clients are limited. The clients' computational capability was modeled as how many data samples could be processed every second to update the model, which can fluctuate because of the computational loads of other applications working on the clients. Therefore, as with the calculation for communication capability, the computation capability was sampled in each round from a truncated normal distribution in (11), where $\mu = \delta_k, \sigma^2 = \delta_k^\eta, a = \delta_k - \sigma, b = \delta_k + \sigma$, at the same time as the throughput sampling. The average computational capability of each client, δ_k, was randomly determined within the range of 10–100 at the beginning of the simulations.

The latency for model updating t_k^{UD} and uploading t_k^{UL} of client k were calculated by the following equations with the above-mentioned sampled throughput θ_k^{tmp} and computational capability δ_k^{tmp}.

$$t_k^{\mathrm{UD}} = D_k/\delta_k^{\mathrm{tmp}}, \tag{13}$$

$$t_k^{\mathrm{UL}} = M/\theta_k^{\mathrm{tmp}}, \tag{14}$$

where D_k denotes the number of data samples stored in client k, and M denotes the data size of the ML model. Each client transmits t_k^{UD} and t_k^{UL} to the server in the `Resource Request` step.

The other hyperparameters were set as follows; $C = 0.1$, the number of clients selected in a round $S_{\mathrm{round}} = 5$. We set $\alpha = 1000$, $\beta = 50$, and $\gamma = 50$ so that T_{inc}, t_k^{UL}, and t_k^{UD} were roughly normalized.

4.3 Simulation Results

The performance of the following three algorithms was evaluated: *Naive MAB-CS*, *Element-wise MAB-CS*, and *Extended FedCS*. *Naive MAB-CS* and *Element-wise MAB-CS* are the proposed MAB-based selection methods detailed in Sect. 3. Specifically, for calculating UCB score, (5) was used in *Naive MAB-CS*, and (9) and (10) were used in *Element-wise MAB-CS*. *Extended FedCS* is an extended algorithm of FedCS detailed in Sect. 3.2.1. In *Extended FedCS*, the expected latency increase for each client is estimated using the moving average of the observed update and upload

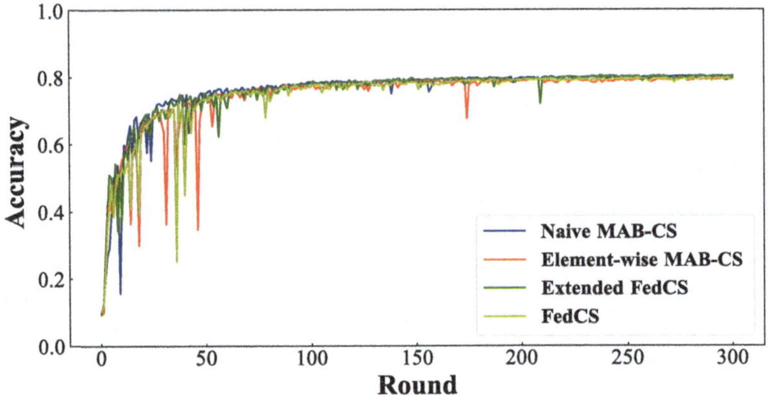

Fig. 1 Effect of client-selection algorithms on the prediction accuracy of trained models when $\eta = 1.5$

latency, and clients with less latency are selected based on the original FedCS algorithm. We evaluated the model accuracy and the reduction in the elapsed times of these three algorithms compared to naive FedCS which uses the last observed latency as the estimated latency of the client.

Figure 1 shows the accuracy-round curve when $\eta = 1.5$. The model test accuracy for all algorithms yielded similar trends. These results suggest that the MAB-based selection does not negatively affect the model accuracy compared to FedCS and its extension that employ the greedy algorithm for the client selection.

Figure 2 illustrates the difference in elapsed time when compared with that of the FedCS as a function of the FL round for two cases: without resource fluctuations and with large resource fluctuations. The y-axis indicates the difference in the elapsed time (i.e., latency reduction):

$$T_{\mathrm{FedCS}} - T_*, \tag{15}$$

where T_{FedCS} and T_* indicate the elapsed times of FedCS and the illustrated algorithm, respectively. As depicted in Fig. 2a, when the resources did not fluctuate, the learning time was longer than that of FedCS for all three algorithms. Among the three algorithms, the increase in the learning time was the largest by *Naive MAB-CS*. We believe that the learning time of the three algorithms was longer than that of FedCS because of unnecessary exploration. When the resources do not fluctuate, the server should only leverage the resource-rich clients without estimating the expected amount of the resources. At the end of the learning process, the difference in the learning time tends not to increase because the exploration of the MAB-based algorithms is already almost complete and exploitation becomes the main task.

In contrast, when resources fluctuate largely, as shown in Fig. 2b, the learning times of the three algorithms were reduced. When $\eta = 1.99$, the learning times of *Naive MAB-CS* and *Element-wise MAB-CS* were significantly shorter than those of FedCS,

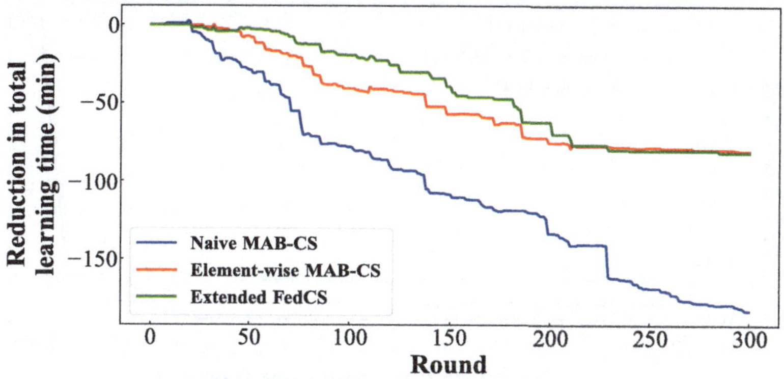

a. No resource fluctuations

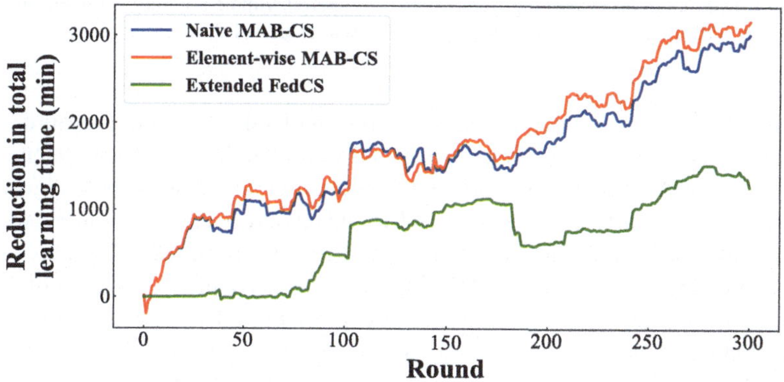

b. $\eta = 1.99$

Fig. 2 Reduction in cumulative learning time at each round. The y-axis displays the difference in the elapsed times at each round for the illustrated algorithms and FedCS. The total learning time at the final round for FedCS was 14415 min in this run

because the MAB-based algorithms can find and leverage resource-rich clients efficiently. Comparing *Naive MAB-CS* and *Element-wise MAB-CS*, the learning time of *Element-wise MAB-CS* was slightly shorter than that of *Naive MAB-CS*. We believe that *Element-wise MAB-CS* improved the convergence of the client-selection algorithm because the UCB scores were not affected by the other arms, a result in contrast with that of *Naive MAB-CS*.

Figure 3 shows latency reduction ratio as a function of η. The dashed lines indicate the ratio when the resource is randomly determined and then fixed (i.e., no resource fluctuation). As also shown in Fig. 2a, when η is small, the reduction rations for all the methods were less than 0, that is, the latency increased. In particular, *Naive MAB-CS* and *Element-wise MAB-CS* increased in latency. This is because of exploration where less selected clients tend to be selected, but frequent exploration is not necessary when η is small. This performance degradation can be mitigated by tuning hyperparameters of UCB score α, β, γ, which is necessary to know in advance how

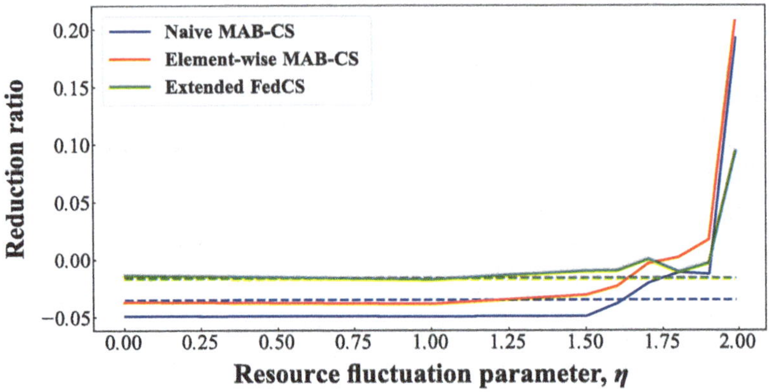

Fig. 3 Ratio of reduction in learning latency when compared with that of FedCS for various η. The dashed lines indicate the reduction ratio when the resources do not fluctuate

much the throughput and computing capability will vary. When the resources fluctuate largely (i.e., η becomes large), these algorithms reduced learning latency, and *Element-wise MAB-CS* achieved the largest reduction ratio among these algorithms. In contrast to the case where η is small, when the resources fluctuate widely, we must select clients based on the average values of their resources, rather than the temporary resource information obtained during resource fluctuation; in such settings, the exploration becomes more important than exploitation. These results demonstrate that in situations where resources vary widely, specifically from 0 to nearly twice the average, MAB-based client selection can reduce learning latency.

Finally, Fig. 4 illustrates the UCB scores for randomly selected 10 clients in *Naive MAB-CS* and *Element-wise MAB-CS* with $\eta = 1.5$. The UCB scores were initially set to a large value and then updated when the client was selected. The reason why some clients kept high UCB scores while other clients were selected is that these clients were not chosen in `Resource Request` step, which selects clients randomly. As increasing rounds, the UCB scores of clients, in particular clients that have relatively large UCB scores and are selected frequently, were converging to certain values. In addition, the UCB scores of the clients with very low UCB scores were less frequently updated, which indicates these clients were selected less frequently.

4.4 Experimental Settings

We implemented FL with client selection algorithms to off-the-shelf devices and conducted experiments to demonstrate its feasibility. Figure 5 depicts the experimental system, which comprised single-board computers (Jetson Nano), MacBooks, and an access point (AP). One of the MacBooks acts as an FL server, and five single-board computers and five MacBooks act as 10 FL clients. The MacBook had a dual

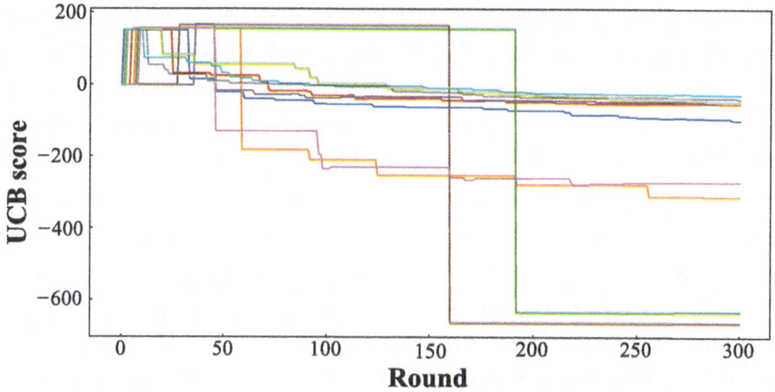

Naive MAB-CS.

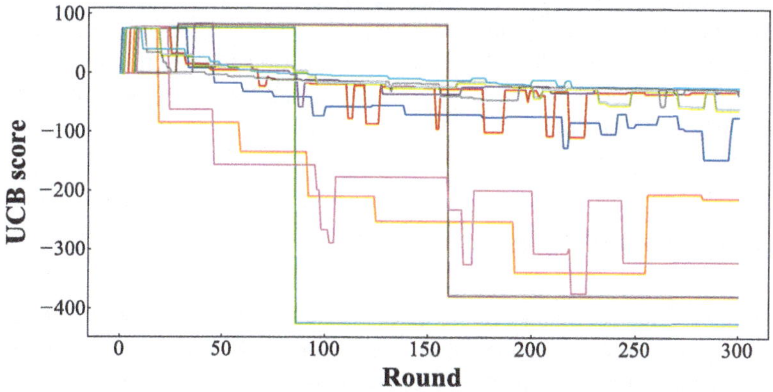

Element-wise MAB-CS.

Fig. 4 Convergence of UCB scores of *Naive MAB-CS* and *Element-wise MAB-CS* when $\eta = 1.5$. Y-axis indicates the UCB score, $f(S, k)$. Each line shows the UCB score of 10 clients out of 100

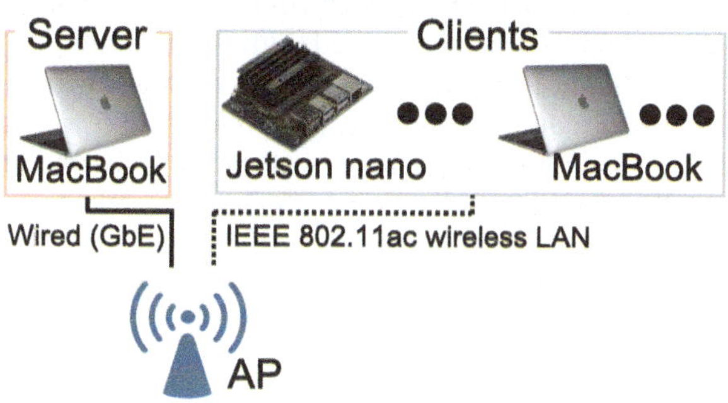

Fig. 5 Experimental setups comprised single-board computers (Jetson Nano), MacBooks, and an AP. The server role MacBook was connected to the AP via Gigabit Ethernet, and others were via IEEE 802.11ac wireless LAN

Table 1 Experimental equipment

Single-board computer	Nvidia Jetson Nano
Laptop	MacBook Pro (Retina 13-inch Late 2012)
AP	NETGEAR Nighthawk X10-AD7200
Wireless LAN	IEEE 802.11ac

2.5 GHz Intel Core i5 processor, and the single-board computer had a quad 1.43 GHz ARM Cortex-A57 MPCore processor. The single-board computer also had 128-core Maxwell GPU that could be used to train neural networks, whereas the MacBook had no GPU for ML. The server was connected to the AP via a wired LAN, and the clients were connected to the AP over an IEEE 802.11ac wireless LAN. The details of the experimental equipment are listed in Table 1.

For evaluating the performance of client selection in an environment where the computation and communication resources fluctuated highly, random pause time (0 s–30 s) was inserted into the `Model Update` and `Scheduled Upload` steps in each round for each client.

4.5 *Experimental Results*

Figure 6 depicts the accuracies of the model trained by FL with four client selection algorithms in the experiment and simulation under similar operating conditions. The four client selection algorithms are Random, FedCS, *Naive MAB-CS*, and *Element-wise MAB-CS*. The experiment and simulation yielded similar prediction accuracy with all four client selection algorithms. The result demonstrates that FL with client selection trains a global model as well in the experimental environment as it does in a computer simulation.

Figure 7 illustrates the reduction in the learning time based on experiments on actual devices when compared with the reduction obtained using FedCS. The time consumption in FedCS for 100 rounds was 71 min. Therefore, the Element-wise MAB-CS reduced the time consumption by 3.5%. The reason for the reduction gain being smaller when compared with that of the simulation results is the significantly smaller number of clients, which degrades selective diversity. We can expect that, in experiments with a large number of clients, FL with client selection will achieve similar performance gains to that in the simulations.

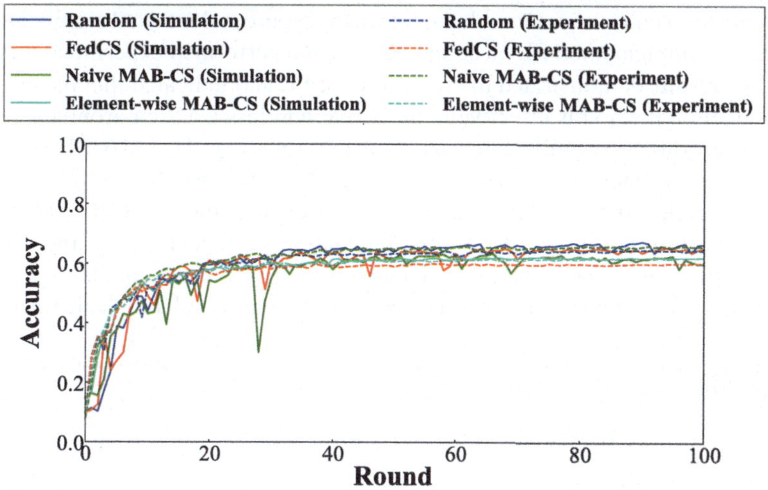

Fig. 6 Prediction accuracies achieved by evaluations under simulated and experimental environments with four client selection algorithms. The solid and dashed lines indicate the simulation and experimental results, respectively

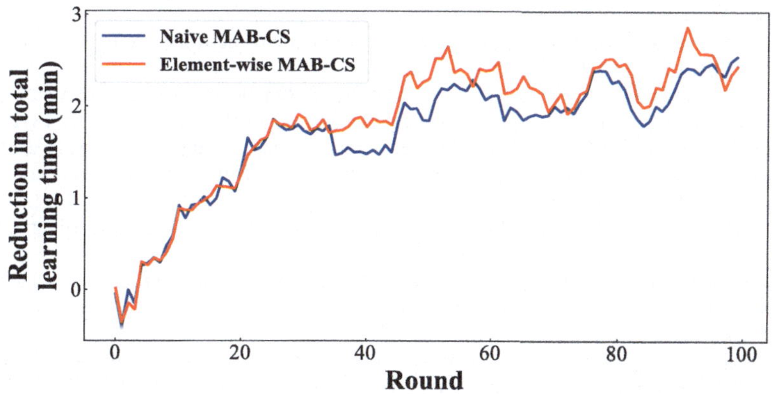

Fig. 7 Reduction in elapsed learning time in the experimental evaluation. The y-axis indicates the difference in elapsed time between those of the illustrated algorithms and FedCS

5 Conclusion

This paper presents an MAB-based client selection method that reduces the FL latency induced by selecting resource-poor clients in scenarios with uncertain and fluctuating resources. The proposed method does not require prior knowledge of the computation and communication resources of clients, and the UCB-based selection balances the exploration and exploitation in the FL client selection. Simulation results demonstrated that the proposed method reduced learning latency (including communication and computation latency) when resources fluctuated significantly. In addition, we implemented FL on actual devices and performed experimental evaluations. The results demonstrated the feasibility of FL and indicated that the proposed client-selection method is effective in the actual wireless LAN environment.

The future directions will include ensuring the efficiency of the MAB-based client-selection method under minor resource fluctuation conditions. For example, choosing the best selection method between the conventional method and the MAB-based method based on the range of resource fluctuations. Another interesting direction will be performing experimental evaluations with hundreds of clients on an actual wireless LAN environment to identify the likely problems at large-scale operations.

Acknowledgements This work was supported in part by JST PRESTO Grant Number JPMJPR2035.

References

1. McMahan B, Moore E, Ramage D, Hampson S, y Arcas BA, (2017) Communication-efficient learning of deep networks from decentralized data. In: Proceeding of the AISTATS, Fort Lauderdale, FL, USA, pp 1273–1282
2. Park J, Samarakoon S, Bennis M, Debbah M (2019) Wireless network intelligence at the edge. Proc. IEEE 107(11):2204–2239
3. Yang T, Andrew G, Eichner H, Sun H, Li W, Kong N, Ramage D, Beaufays F (2018) Applied federated learning: Improving Google keyboard query suggestions, pp 1–9. arXiv:1812.02903
4. Hu B, Gao Y, Liu L, Ma H (2018) Federated region-learning: an edge computing based framework for urban environment sensing. In: Proceeding of the IEEE global communications conference (GLOBECOM), pp 1–7
5. Li L, Fan Y, Tse M, Lin K-Y (2020) A review of applications in federated learning. Comput & Ind Eng 149:106854
6. Lin Y, Han S, Mao H, Wang Y, Dally WJ (2018) Deep gradient compression: reducing the communication bandwidth for distributed training. In: Proceedings of the ICLR, Vancouver, BC, Canada, pp 1–14
7. Bonawitz K, Eichner H, Grieskamp W, Huba D, Ingerman A, Ivanov V, Kiddon C, Konecný J, Mazzocchi S, McMahan HB, Overveldt TV, Petrou D, Ramage D, Roselander J (2019) Towards federated learning at scale: system design, pp 1–15. arXiv:1902.01046
8. Li T, Sahu AK, Talwalkar A, Smith V (2020) Federated learning: Challenges, methods, and future directions. IEEE Signal Process Mag 37(3):50–60
9. Kairouz P, McMahan HB, Avent B, Bellet A, Bennis M, Bhagoji AN, Bonawitz K, Charles Z, Cormode G, Cummings R et al (2021) Advances and open problems in federated learning. Found Trends Mach Learn 14(1)
10. Lim WYB, Luong NC, Hoang DT, Jiao Y, Liang YC, Yang Q, Niyato D, Miao C (2020) Federated learning in mobile edge networks: a comprehensive survey. IEEE Commun Surveys Tuts 22(3):2031–2063
11. Chen J, Monga R, Bengio S, Józefowicz R (2017) Revisiting distributed synchronous SGD, pp 1–10. arXiv:1604.00981
12. Mills J, Hu J, Min G (2020) Communication-efficient federated learning for wireless edge intelligence in IoT. IEEE Internet Things J 7(7):5986–5994
13. Zhu H, Jin Y (2020) Multi-objective evolutionary federated learning. IEEE Trans Neural Netw Learn Syst 31(4):1310–1322

14. Sattler F, Wiedemann S, Müller KR, Samek W (2020) Robust and communication-efficient federated learning from non-i.i.d. data. IEEE Trans Neural Netw Learn Syst 31(9):3400–3413
15. Nishio T, Yonetani R (2019) Client selection for federated learning with heterogeneous resources in mobile edge. In: Proceeding of the IEEE ICC, Shanghai, China, pp 1–7
16. Dhakal S, Prakash S, Yona Y, Talwar S, Himayat N (2019) Coded federated learning. In: IEEE globecom workshops, HI, USA, Waikoloa, pp 1–6
17. Yoshida N, Nishio T, Morikura M, Yamamoto K, Yonetani R (2020) Hybrid-FL for wireless networks: cooperative learning mechanism using non-IID data. In: Proceeding of the IEEE ICC, Dublin, Ireland, pp 1–7
18. Yang HH, Arafa A, Quek TQS, Poor HV (2020) Age-based scheduling policy for federated learning in mobile edge networks. In: Proceeding of the IEEE ICASSP, Barcelona, Spain, pp 8743–8747
19. Chen M, Yang Z, Saad W, Yin C, Poor HV, Cui S (2020) A joint learning and communications framework for federated learning over wireless networks. IEEE Trans Wireless Commun 1–14
20. Auer P, Cesa-Bianchi N, Fischer P (2002) Finite-time analysis of the multiarmed bandit problem. Mach Learn 47(2–3):235–256
21. Li F, Yu D, Yang H, Yu J, Karl H, Cheng X (2020) Multi-armed-bandit-based spectrum scheduling algorithms in wireless networks: a survey. IEEE Wireless Commun 27(1):24–30
22. Zhao P, Tian H, Chen K, Fan S, Nie G (2020) Context-aware TDD configuration and resource allocation for mobile edge computing. IEEE Trans Commun 68(2):1118–1131
23. Yoshida N, Nishio T, Morikura M, Yamamoto K (2020) MAB-based client selection for federated learning with uncertain resources in mobile networks. In: Proceeding of the IEEE globecom workshops (GC Wkshps), pp 1–6
24. Cho YJ, Gupta S, Joshi G, Yağan O (2020) Bandit-based communication-efficient client selection strategies for federated learning. In: Proceeding of the Asilomar conference on signals, systems, and computers, pp 1066–1069
25. Xia W, Quek TQS, Guo K, Wen W, Yang HH, Zhu H (2020) Multi-armed bandit-based client scheduling for federated learning. IEEE Trans Wireless Commun 19(11):7108–7123
26. Sesia S, Baker M, Toufik I (2011) LTE, The UMTS long term evolution: from theory to practice. Wiley
27. Krizhevsky A, Hinton G (2009) Learning multiple layers of features from tiny images. University of Toronto, Technical report
28. Konecný J, McMahan HB, Yu FX, Richtárik P, Suresh AT, Bacon D (2016) Federated learning: strategies for improving communication efficiency. In: Proceeding of the NIPS Workshops, Barcelona, Spain
29. Series M (2009) Guidelines for evaluation of radio interface technologies for IMT-Advanced. Report ITU-R M.2135-1
30. Akdeniz MR, Liu Y, Samimi MK, Sun S, Rangan S, Rappaport TS, Erkip E (2014) Millimeter wave channel modeling and cellular capacity evaluation. IEEE J Sel Areas Commun 32(6):1164–1179

Index

R. Shinkuma et al. (eds.), *Advances in Engineering and Information Science Toward Smart City and Beyond*, Engineering Cyber-Physical Systems and Critical Infrastructures 5, https://doi.org/10.1007/978-3-031-29301-6

The manufacturer's authorised representative in the EU is Springer Nature Customer Service Centre GmbH, Europaplatz 3, 69115 Heidelberg, Germany. If you have any concerns regarding our products, please contact ProductSafety@springernature.com

Printed and bound by CPI Group (UK) Ltd, Croydon, CR0 4YY
30/09/2026
02225303-0009